REVIEW OF THE
Formaldehyde Assessment
IN THE National Toxicology Program 12th Report on Carcinogens

Committee to Review the Formaldehyde Assessment in the
National Toxicology Program 12th Report on Carcinogens

Board on Environmental Studies and Toxicology

Division on Earth and Life Studies

NATIONAL RESEARCH COUNCIL
OF THE NATIONAL ACADEMIES

THE NATIONAL ACADEMIES PRESS
Washington, D.C.
www.nap.edu

THE NATIONAL ACADEMIES PRESS   500 Fifth Street, NW   Washington, DC 20001

NOTICE: The project that is the subject of this report was approved by the Governing Board of the National Research Council, whose members are drawn from the councils of the National Academy of Sciences, the National Academy of Engineering, and the Institute of Medicine. The members of the committee responsible for the report were chosen for their special competences and with regard for appropriate balance.

This project was supported by Contract HHSP233201200024C between the National Academy of Sciences and the the Department of Health and Human Services. Any opinions, findings, conclusions, or recommendations expressed in this publication are those of the authors and do not necessarily reflect the view of the organizations or agencies that provided support for this project.

International Standard Book Number-13: 978-0-309-31227-1
International Standard Book Number-10: 0-309-31227-2

Additional copies of this report are available for sale from the National Academies Press, 500 Fifth Street, NW, Keck 360, Washington, DC 20001; (800) 624-6242 or (202) 334-3313; http://www.nap.edu/.

Copyright 2014 by the National Academy of Sciences. All rights reserved.

Printed in the United States of America

THE NATIONAL ACADEMIES
Advisers to the Nation on Science, Engineering, and Medicine

The **National Academy of Sciences** is a private, nonprofit, self-perpetuating society of distinguished scholars engaged in scientific and engineering research, dedicated to the furtherance of science and technology and to their use for the general welfare. Upon the authority of the charter granted to it by the Congress in 1863, the Academy has a mandate that requires it to advise the federal government on scientific and technical matters. Dr. Ralph J. Cicerone is president of the National Academy of Sciences.

The **National Academy of Engineering** was established in 1964, under the charter of the National Academy of Sciences, as a parallel organization of outstanding engineers. It is autonomous in its administration and in the selection of its members, sharing with the National Academy of Sciences the responsibility for advising the federal government. The National Academy of Engineering also sponsors engineering programs aimed at meeting national needs, encourages education and research, and recognizes the superior achievements of engineers. Dr. C. D. Mote, Jr., is president of the National Academy of Engineering.

The **Institute of Medicine** was established in 1970 by the National Academy of Sciences to secure the services of eminent members of appropriate professions in the examination of policy matters pertaining to the health of the public. The Institute acts under the responsibility given to the National Academy of Sciences by its congressional charter to be an adviser to the federal government and, upon its own initiative, to identify issues of medical care, research, and education. Dr. Victor J. Dzau is president of the Institute of Medicine.

The **National Research Council** was organized by the National Academy of Sciences in 1916 to associate the broad community of science and technology with the Academy's purposes of furthering knowledge and advising the federal government. Functioning in accordance with general policies determined by the Academy, the Council has become the principal operating agency of both the National Academy of Sciences and the National Academy of Engineering in providing services to the government, the public, and the scientific and engineering communities. The Council is administered jointly by both Academies and the Institute of Medicine. Dr. Ralph J. Cicerone and Dr. C. D. Mote, Jr., are chair and vice chair, respectively, of the National Research Council.

www.national-academies.org

COMMITTEE TO REVIEW THE FORMALDEHYDE
ASSESSMENT IN THE NTP 12TH RoC

Members

ALFRED O. BERG (*Chair*), University of Washington, Seattle
JOHN C. BAILAR III, University of Chicago (retired), Mitchellville, MD
A. JAY GANDOLFI, University of Arizona (retired), Tucson
DAVID KRIEBEL, University of Massachusetts, Lowell
JOHN B. MORRIS, University of Connecticut, Storrs
KENT E. PINKERTON, University of California, Davis
IVAN RUSYN, University of North Carolina, Chapel Hill
TOSHIHIRO SHIODA, Harvard Medical School and Massachusetts General Hospital, Boston
THOMAS J. SMITH, Harvard School of Public Health (retired), Boston, MA
MEIR WETZLER, Roswell Park Cancer Institute, University at Buffalo, State University of New York
LAUREN ZEISE, California Environmental Protection Agency, Oakland
PATRICK ZWEIDLER-MCKAY, The University of Texas M D Anderson Cancer Center, Houston

Staff

HEIDI MURRAY-SMITH, Project Director
KERI STOEVER, Research Associate
NORMAN GROSSBLATT, Senior Editor
MIRSADA KARALIC-LONCAREVIC, Manager, Technical Information Center
RADIAH ROSE, Manager, Editorial Projects
RICARDO PAYNE, Program Coordinator

Sponsor

US DEPARTMENT OF HEALTH AND HUMAN SERVICES

BOARD ON ENVIRONMENTAL STUDIES AND TOXICOLOGY[1]

Members

ROGENE F. HENDERSON (*Chair*), Lovelace Respiratory Research Institute, Albuquerque, NM
PRAVEEN AMAR, Clean Air Task Force, Boston, MA
RICHARD A. BECKER, American Chemistry Council, Washington, DC
MICHAEL J. BRADLEY, M.J. Bradley & Associates, Concord, MA
JONATHAN Z. CANNON, University of Virginia, Charlottesville
GAIL CHARNLEY, HealthRisk Strategies, Washington, DC
DAVID C. DORMAN, Department of Molecular Biomedical Sciences, Raleigh, NC
CHARLES T. DRISCOLL, JR., Syracuse University, New York
WILLIAM H. FARLAND, Colorado State University, Fort Collins, CO
LYNN R. GOLDMAN, George Washington University, Washington, DC
LINDA E. GREER, Natural Resources Defense Council, Washington, DC
WILLIAM E. HALPERIN, University of Medicine and Dentistry of New Jersey, Newark
STEVEN P. HAMBURG, Environmental Defense Fund, New York, NY
ROBERT A. HIATT, University of California, San Francisco
PHILIP K. HOPKE, Clarkson University, Potsdam, NY
SAMUEL KACEW, University of Ottawa, Ontario
H. SCOTT MATTHEWS, Carnegie Mellon University, Pittsburgh, PA
THOMAS E. MCKONE, University of California, Berkeley
TERRY L. MEDLEY, E.I. du Pont de Nemours & Company, Wilmington, DE
JANA MILFORD, University of Colorado at Boulder, Boulder
MARK A. RATNER, Northwestern University, Evanston, IL
JOAN B. ROSE, Michigan State University, East Lansing, MI
GINA M. SOLOMON, California Environmental Protection Agency, Sacramento, CA
PETER S. THORNE, University of Iowa, Iowa City, IA
DOMINIC M. DI TORO, University of Delaware Newark, DE
JOYCE S. TSUJI, Exponent, Bellevue, WA

Senior Staff

JAMES J. REISA, Director
DAVID J. POLICANSKY, Scholar
RAYMOND A. WASSEL, Senior Program Officer for Environmental Studies
ELLEN K. MANTUS, Senior Program Officer for Risk Analysis
SUSAN N.J. MARTEL, Senior Program Officer for Toxicology
MIRSADA KARALIC-LONCAREVIC, Manager, Technical Information Center
RADIAH ROSE, Manager, Editorial Projects

---

[1]This study was planned, overseen, and supported by the Board on Environmental Studies and Toxicology.

OTHER REPORTS OF THE
BOARD ON ENVIRONMENTAL STUDIES AND TOXICOLOGY

Review of the Styrene Assessment in the National Toxicology Program 12th Report on Carcinogens (2014)
Review of EPA's Integrated Risk Information System (IRIS) Process (2014)
Review of the Environmental Protection Agency's State-of-the-Science Evaluation of Nonmonotonic Dose–Response Relationships as They Apply to Endocrine Disruptors (2014)
Assessing Risks to Endangered and Threatened Species from Pesticides (2013)
Science for Environmental Protection: The Road Ahead (2012)
Exposure Science in the 21st Century: A Vision and A Strategy (2012)
A Research Strategy for Environmental, Health, and Safety Aspects of Engineered Nanomaterials (2012)
Macondo Well–Deepwater Horizon Blowout: Lessons for Improving Offshore Drilling Safety (2012)
Feasibility of Using Mycoherbicides for Controlling Illicit Drug Crops (2011)
Improving Health in the United States: The Role of Health Impact Assessment (2011)
A Risk-Characterization Framework for Decision-Making at the Food and Drug Administration (2011)
Review of the Environmental Protection Agency's Draft IRIS Assessment of Formaldehyde (2011)
Toxicity-Pathway-Based Risk Assessment: Preparing for Paradigm Change (2010)
The Use of Title 42 Authority at the U.S. Environmental Protection Agency (2010)
Review of the Environmental Protection Agency's Draft IRIS Assessment of Tetrachloroethylene (2010)
Hidden Costs of Energy: Unpriced Consequences of Energy Production and Use (2009)
Contaminated Water Supplies at Camp Lejeune—Assessing Potential Health Effects (2009)
Review of the Federal Strategy for Nanotechnology-Related Environmental, Health, and Safety Research (2009)
Science and Decisions: Advancing Risk Assessment (2009)
Phthalates and Cumulative Risk Assessment: The Tasks Ahead (2008)
Estimating Mortality Risk Reduction and Economic Benefits from Controlling Ozone Air Pollution (2008)
Respiratory Diseases Research at NIOSH (2008)
Evaluating Research Efficiency in the U.S. Environmental Protection Agency (2008)
Hydrology, Ecology, and Fishes of the Klamath River Basin (2008)
Applications of Toxicogenomic Technologies to Predictive Toxicology and Risk Assessment (2007)
Models in Environmental Regulatory Decision Making (2007)
Toxicity Testing in the Twenty-first Century: A Vision and a Strategy (2007)
Sediment Dredging at Superfund Megasites: Assessing the Effectiveness (2007)
Environmental Impacts of Wind-Energy Projects (2007)
Scientific Review of the Proposed Risk Assessment Bulletin from the Office of Management and Budget (2007)
Assessing the Human Health Risks of Trichloroethylene: Key Scientific Issues (2006)
New Source Review for Stationary Sources of Air Pollution (2006)
Human Biomonitoring for Environmental Chemicals (2006)

Health Risks from Dioxin and Related Compounds: Evaluation of the EPA Reassessment (2006)
Fluoride in Drinking Water: A Scientific Review of EPA's Standards (2006)
State and Federal Standards for Mobile-Source Emissions (2006)
Superfund and Mining Megasites—Lessons from the Coeur d'Alene River Basin (2005)
Health Implications of Perchlorate Ingestion (2005)
Air Quality Management in the United States (2004)
Endangered and Threatened Species of the Platte River (2004)
Atlantic Salmon in Maine (2004)
Endangered and Threatened Fishes in the Klamath River Basin (2004)
Cumulative Environmental Effects of Alaska North Slope Oil and Gas Development (2003)
Estimating the Public Health Benefits of Proposed Air Pollution Regulations (2002)
Biosolids Applied to Land: Advancing Standards and Practices (2002)
The Airliner Cabin Environment and Health of Passengers and Crew (2002)
Arsenic in Drinking Water: 2001 Update (2001)
Evaluating Vehicle Emissions Inspection and Maintenance Programs (2001)
Compensating for Wetland Losses Under the Clean Water Act (2001)
A Risk-Management Strategy for PCB-Contaminated Sediments (2001)
Acute Exposure Guideline Levels for Selected Airborne Chemicals (seventeen volumes, 2000-2014)
Toxicological Effects of Methylmercury (2000)
Strengthening Science at the U.S. Environmental Protection Agency (2000)
Scientific Frontiers in Developmental Toxicology and Risk Assessment (2000)
Ecological Indicators for the Nation (2000)
Waste Incineration and Public Health (2000)
Hormonally Active Agents in the Environment (1999)
Research Priorities for Airborne Particulate Matter (four volumes, 1998-2004)
The National Research Council's Committee on Toxicology: The First 50 Years (1997)
Carcinogens and Anticarcinogens in the Human Diet (1996)
Upstream: Salmon and Society in the Pacific Northwest (1996)
Science and the Endangered Species Act (1995)
Wetlands: Characteristics and Boundaries (1995)
Biologic Markers (five volumes, 1989-1995)
Science and Judgment in Risk Assessment (1994)
Pesticides in the Diets of Infants and Children (1993)
Dolphins and the Tuna Industry (1992)
Science and the National Parks (1992)
Human Exposure Assessment for Airborne Pollutants (1991)
Rethinking the Ozone Problem in Urban and Regional Air Pollution (1991)
Decline of the Sea Turtles (1990)

Copies of these reports may be ordered from the National Academies Press
(800) 624-6242 or (202) 334-3313
www.nap.edu

Preface

In 1981, The National Toxicology Program (NTP) first listed formaldehyde in the 2nd Report on Carcinogens (RoC) as "reasonably anticipated to be a human carcinogen". In 2011, NTP upgraded the listing of formaldehyde in the 12th RoC to "known to be a human carcinogen". Following the new listing, Congress directed the Department of Health and Human Services to arrange for the National Academy of Sciences (NAS) to independently review formaldehyde's substance profile and listing in the 12th RoC (112th Congress, 1st Session; Public Law 112-74). This report presents the findings and conclusions of the committee formed in response to the congressional request.

To address its statement of task, the committee first conducted a peer review of the formaldehyde substance profile and listing in the 12th RoC. It considered literature available to NTP up to the publication of the 12th RoC (that is, literature published by June 10, 2011). The committee then conducted an independent assessment of formaldehyde and made a listing recommendation using the RoC listing criteria. In its independent assessment, the committee examined evidence published both before and after the publication of the 12th RoC. It considered presentations heard during its open-session meeting, comments submitted from the general public, and abstracts presented during conferences. It reviewed reports published by other authoritative bodies, and it examined primary literature, reviews, and meta-analyses that were publicly available in the peer-reviewed literature.

This report has been reviewed in draft form by persons chosen for their diverse perspectives and technical expertise in accordance with procedures approved by the National Research Council Report Review Committee. The purpose of the independent review is to provide candid and critical comments that will assist the institution in making its published report as sound as possible and to ensure that the report meets institutional standards of objectivity, evidence, and responsiveness to the study charge. The review comments and draft manuscript remain confidential to protect the integrity of the deliberative process. We thank the following for their review of the report: Hugh Barton, Pfizer, Inc.; Harvey Checkoway, University of California, San Diego; David C. Dorman, North Carolina State University; Rogene F. Henderson; Lovelace Respiratory

Research Institute; Charles G. Mullighan, St. Jude Children's Research Hospital; Neil Pearce, London School of Hygiene and Tropical Medicine; Elizabeth A. Platz, Johns Hopkins University; Joseph V. Rodricks, ENVIRON; Jonathan M. Samet, University of Southern California; Noah S. Seixas, University of Washington School of Public Health and Community Medicine; Michael J. Thirman, The University of Chicago Medicine; and Gerald N. Wogan, Massachusetts Institute of Technology.

Although the reviewers listed above have provided many constructive comments and suggestions, they were not asked to endorse the conclusions or recommendations, nor did they see the final draft of the report before its release. The review of the report was overseen by the review coordinator, Kenneth Ramos, University of Arizona, and the review monitor, Donald Mattison, Risk Sciences International. Appointed by the National Research Council, they were responsible for making certain that an independent examination of the report was carried out in accordance with institutional procedures and that all review comments were carefully considered. Responsibility for the final content of the report rests entirely with the committee and the institution.

The committee gratefully acknowledges Dr. Wanda Jones, U.S. Department of Health and Human Services, and Dr. John Bucher, National Toxicology Program, for making presentations to the committee. The committee appreciates all who supplied written documents or views to the committee during its open public session and throughout the study process.

The committee is also grateful for the assistance of the National Research Council staff in preparing this report. Staff members who contributed to the effort are Heidi Murray-Smith, project director; Ellen Mantus, senior program officer; Keri Stoever, research associate; James Reisa, director of the Board on Environmental Studies and Toxicology; Norman Grossblatt, senior editor; Mirsada Karalic-Loncarevic, manager of the Technical Information Center; Radiah Rose, manager of editorial projects; and Ricardo Payne, program coordinator.

I especially thank the members of the committee for contributing their outstanding expertise, scientific focus, meticulous attention to detail, tireless hard work, and consistent good humor throughout the development of this report.

Alfred O. Berg, *Chair*
Committee to Review the
Formaldehyde Assessment in the
National Toxicology Program
12th Report on Carcinogens

Contents

SUMMARY 3

1 INTRODUCTION 20
The Report on Carcinogens, 20
Formaldehyde and the Report on Carcinogens, 25
The Committee's Task, 26
The Committee's Approach, 26
Organization of the Report, 30
References, 30

**2 REVIEW OF THE FORMALDEHYDE PROFILE IN
THE NATIONAL TOXICOLOGY PROGRAM 12th
REPORT ON CARCINOGENS** 33
Carcinogenicity, 34
Properties, 53
Use, 53
Production, 54
Exposure, 54
Regulations and Guidelines, 54
Review of NTP'S Literature-Search Methods, 55
Suggested Revisions for Future Editions of the Formaldehyde Listing
   in the Report on Carginogens, 57
Summary and Conclusions, 57
References, 60

3 INDEPENDENT ASSESSMENT OF FORMALDEHYDE 66
Cancer Studies in Humans, 67
Cancer Studies in Experimental Animals, 122
Toxicokinetics, 129
Mechanisms of Carcinogenesis, 136
Summary of Evidence, 163
Conclusions and Listing Recommendation, 165
References, 167

APPENDIXES

A BIOGRAPHICAL INFORMATION ON THE COMMITTEE
   TO REVIEW THE FORMALDEHYDE ASSESSMENT IN THE
   NTP 12th ROC 179

B STATEMENT OF TASK FOR THE COMMITTEE TO REVIEW
   THE FORMALDEHYDE ASSESSMENT IN THE NTP 12th ROC 184

C EXPOSURE ASSESSMENT IN EPIDEMIOLOGIC
   CARCINOGENICITY STUDIES 185

D LITERATURE-SEARCH STRATEGIES COMPLETED IN
   SUPPORT OF THE COMMITTEE'S INDEPENDENT
   ASSESSMENT OF FORMALDEHYDE 199

E GENOTOXICITY AND MUTAGENICITY SUMMARY TABLES 207

BOXES, FIGURES, AND TABLES

BOXES

1-1 Congressional Language Mandating the Report on Carcinogens, 22
1-2 Listing Criteria for the Report on Carcinogens, 23
3-1 Guidance from Various Agencies on the Use of Mechanistic and Other Relevant Data, 139
D-1 Exclusion Criteria and Search Strategy for Human Studies, 200
D-2 Exclusion Criteria and Search Strategy for Experimental-Animal Studies, 202
D-3 Exclusion Criteria and Search Strategy for Genotoxicity and Mutagenicity Mechanisms of Carcinogenesis, 203
D-4 Exclusion Criteria and Search Strategy for Immune Effects, 205

FIGURES

1-1 Schematic of the review process for the 12th Report on Carcinogens, 24
3-1 Summary of strong and moderately strong studies of formaldehyde and lymphohematopoietic cancers, 118
3-2 Summary of key findings from all studies that reported associations between formaldehyde and myeloid leukemia, 119
3-3 Schematic representation of the structure of the nasal mucosa of the respiratory epithelium and follicle-associated epithelium, 134
3-4 Model-based estimates of exogenous formaldehyde concentration in nasal tissues during inhalation exposure to 6 ppm formaldehyde, 135
D-1 Literature tree for human studies search, 201
D-2 Literature tree for experimental-animal studies search, 202
D-3 Literature tree for genotoxicity search, 204
D-4 Literature tree for immune-effects search, 206

TABLES

1-1 Documents Pertaining to Formaldehyde That Were Available to or Written by NTP, 27
2-1 Topic-Specific Search Terms Used in NTP's Database Searches, 56
2-2 Suggested Revisions for the Formaldehyde Substance Profile and Background Document in Future Editions of the Report on Carcinogens, 58
3-1 Criteria Used to Assess Epidemiologic Studies for Hazard Assessment, 69
3-2 Description of Epidemiologic Studies Reviewed by the Committee, 70
3-3 Studies of Nasopharyngeal Cancer and Formaldehyde Exposure, 86
3-4 Studies of Sinonasal Cancer and Formaldehyde Exposure, 93
3-5 Lymphohematopoietic Cancers: Industrial Workers, 103
3-6 Lymphohematopoietic Cancers: Funeral Workers, Embalmers, Pathologists, and Anatomists, 111
3-7 Other Cancer Sites, 120
3-8 Studies of Low Power for Detecting Malignancies, 123
3-9 Nasal Squamous-Cell Carcinoma in Long-Term Inhalation Studies of Formaldehyde, 125
3-10 Summary of Published Studies on the Genotoxic and Mutagenic Effects of Formaldehyde in Test Systems and Organisms, 141
3-11 Recent Studies of Hematologic Effects of Formaldehyde, 148
3-12 Studies Grouped by Hematologic Effects, 153
3-13 Transcriptomal Profiling Studies, 158
C-1 Distinctions between Different Levels of Exposure, 188
C-2 Information Used to Evaluate Exposure Assessment Components of Epidemiologic Studies in Chapters 2 and 3, 196
E-1 DNA Adducts, 208
E-2 DNA–DNA Cross-Links, 209
E-3 DNA–Protein Cross-Links, 210
E-4 DNA Strand Breaks, 212
E-5 Mutations, 214
E-6 Sister-Chromatid Exchanges, 215
E-7 Micronuclei, 217
E-8 Chromosomal Aberrations, 219

REVIEW OF THE
Formaldehyde Assessment
IN THE National Toxicology Program
12th Report on Carcinogens

Summary

As part of the 2012 Consolidated Appropriations Act (112th Congress, 1st Session; Public Law 112-74), Congress directed the Department of Health and Human Services to arrange for the National Academy of Sciences to carry out an independent review of the formaldehyde assessment in the National Toxicology Program (NTP) 12th Report on Carcinogens (RoC).[1] In response, the Academy's National Research Council (NRC) convened an expert committee that has prepared this report.

THE COMMITTEE'S APPROACH

The NRC Committee to Review the Formaldehyde Assessment in the NTP 12th RoC approached its statement of task by first conducting a review of the substance profile for formaldehyde as presented in the 12th RoC. It considered literature published by June 10, 2011 (reflecting the date of publication of the 12th RoC), and it organized its review on the basis of the headings and subheadings of the substance profile. The committee then conducted its own independent assessment of the formaldehyde literature, extending its review to include literature through November 8, 2013, and concluding with its own listing recommendation for formaldehyde.

The committee noted that the assessment of chemicals for the purposes of listing in the RoC constitutes a hazard assessment, not a risk assessment. A hazard assessment focuses on the identification of substances that may pose a hazard to human health, and it "makes a classification regarding toxicity, for example, whether a chemical is 'carcinogenic to humans' or 'likely to be'."[2] A risk assessment focuses on the likely degree of damage and requires much more information, including completion of a hazard identification, dose–response analy-

---

[1]NTP (National Toxicology Program). 2011. Formaldehyde. Pp. 195-205 in Report on Carcinogens, 12th Ed. U.S. Department of Health and Human Services, Public Health Service, National Toxicology Program, Research Triangle Park, NC [online]. Available: http://ntp.niehs.nih.gov/ntp/roc/twelfth/profiles/formaldehyde.pdf.

[2]NRC (National Research Council). 2009. Science and Decision: Advancing Risk Assessment. Washington, DC: National Academies Press.

sis, exposure quantification, and characterization of risk. The committee thus approached its assessment of formaldehyde as an evaluation of hazard, not risk. It evaluated measures of association in a population (such as risk ratios, odds ratios, and incidence ratios) from epidemiology studies to inform its assessment of formaldehyde, but it did not identify exposure scenarios that could pose cancer risk as part of a full risk assessment.

The committee examined the 2011 NRC report, *Review of the Environmental Protection Agency (EPA) Draft IRIS (Integrated Risk Information System) Assessment of Formaldehyde*.[3] Although the present report and the 2011 report both focused on formaldehyde, the two committees had different statements of task. The Committee to Review EPA's Draft IRIS Assessment of Formaldehyde was asked to "conduct an independent scientific review of [EPA's] draft human health assessment of formaldehyde for [IRIS]." It was also asked to address specific questions related to EPA's inhalation reference concentration (RfC) for noncancer health effects and its risk estimate for carcinogenicity. That committee assessed how well the narrative presented in the draft IRIS assessment supported the IRIS assessment's conclusions regarding health effects. That committee did not conduct its own literature search, review all relevant evidence, systematically formulate its own conclusions regarding causality, or recommend values for the RfC and unit risk. In contrast, the committee that wrote the present report was asked to identify relevant peer-reviewed literature, document its decisions regarding inclusion or exclusion of the literature, apply NTP's RoC listing criteria, and make an independent listing recommendation for formaldehyde.

The two projects were also different because of inherent differences between EPA's IRIS assessments and NTP's RoC. IRIS assessments are comprehensive human health assessments that evaluate cancer and noncancer end points and include hazard and dose-response assessments that are used to derive toxicity values (that is, reference values and unit risk values), whereas NTP qualitatively weighs evidence of carcinogenicity and compiles lists of substances that it classifies as known human carcinogens or reasonably anticipated human carcinogens to produce the biennial RoC. Because of those differences, the committee cautions readers against making direct comparisons between the two reports.

THE NATIONAL TOXICOLOGY PROGRAM AND FORMALDEHYDE

NTP is an interagency program involving the National Institutes of Health's National Institute of Environmental Health Sciences (the administrative lead), the Centers for Disease Control and Prevention's National Institute for

---

[3]NRC (National Research Council). 2011. Review of the Environmental Protection Agency's Draft IRIS Assessment of Formaldehyde. Washington, DC: National Academies Press.

Occupational Safety and Health, and the Food and Drug Administration's National Center for Toxicological Research. Since 1980, NTP has published the RoC, which is a cumulative summary of substances that have been nominated for review and judged to meet two conditions. The first condition is that a significant number of people living in the United States are exposed to the substance of interest. The second condition is that there is judged to be evidence that the substance of interest is either known to be a human carcinogen or reasonably anticipated to be a human carcinogen on the basis of NTP's established listing criteria.

In the late 1970s and early 1980s, NTP assessed the potential carcinogenicity of formaldehyde, and the substance was listed as "reasonably anticipated to be a human carcinogen" in the 2nd RoC (1981). Three decades later, NTP reassessed formaldehyde and upgraded its listing to "known to be a human carcinogen" in the 12th RoC (2011). Formaldehyde is a substance of interest because many people in the United States are exposed. Exposure can occur from environmental sources (for example, combustion processes, building materials, and tobacco smoke) or in occupational settings (for example, the furniture, textile, and construction industries). Formaldehyde exposure also has endogenous sources—it is produced in humans intracellularly as a component of the one-carbon pool intermediary metabolism pathway. Scientists have studied formaldehyde for decades to determine whether exogenous formaldehyde exposure may be associated with cancer in humans. Much of the focus has been on cancers of the upper respiratory tract because those tissues were thought to be the most biologically plausible targets. However, there is increasing interest in a potential relationship between formaldehyde exposure and some lymphohematopoietic cancers (for example, leukemia).

The Report on Carcinogens Listing Criteria

The committee's assessment of formaldehyde was guided by the RoC listing criteria.[4] A substance can be classified in the RoC as "reasonably anticipated to be a human carcinogen" if at least one of the following criteria is fulfilled:

- "There is limited evidence of carcinogenicity from studies in humans, which indicates that causal interpretation is credible, but that alternative explanations, such as chance, bias, or confounding factors, could not adequately be excluded."

---

[4]NTP (National Toxicology Program). 2010. Report on Carcinogens Background Document for Formaldehyde, January 22, 2010. U.S. Department of Health and Human Services, Public Health Service, National Toxicology Program, Research Triangle Park, NC [online]. Available: http://ntp.niehs.nih.gov/ntp/roc/twelfth/2009/November/Formaldehyde_BD_Final.pdf.

- "There is sufficient evidence of carcinogenicity from studies in experimental animals, which indicates there is an increased incidence of malignant and/or a combination of malignant and benign tumors (1) in multiple species or at multiple tissue sites, or (2) by multiple routes of exposure, or (3) to an unusual degree with regard to incidence, site, or type of tumor, or age at onset."
- "There is less than sufficient evidence of carcinogenicity in humans or laboratory animals; however, the agent, substance, or mixture belongs to a well-defined, structurally related class of substances whose members are listed in a previous Report on Carcinogens as either known to be a human carcinogen or reasonably anticipated to be a human carcinogen, or there is convincing relevant information that the agent acts through mechanisms indicating it would likely cause cancer in humans."

A substance can be listed in the RoC as "known to be a human carcinogen" if "there is sufficient evidence of carcinogenicity from studies in humans, which indicates a causal relationship between exposure to the agent, substance, or mixture, and human cancer." Sufficient evidence in humans from only one type of cancer is adequate for a substance to be listed in the RoC as "known to be a human carcinogen". Evidence in experimental animals and a known mechanism of action can provide supporting evidence, but that information is not required by the RoC listing criteria in making a listing recommendation that a substance is known to be a human carcinogen.

The committee found the RoC listing criteria to be clear about the information needed to fulfill the criteria of sufficient evidence in experimental animals; however, the type of information needed to meet the RoC listing criteria for limited or sufficient evidence in humans required more interpretation. Therefore, consistent with the RoC listing criteria, the committee used its expert scientific judgment to interpret and apply the listing criteria to the evidence evaluated in Chapters 2 and 3. It established its own set of evaluation attributes and made judgments on the strength of each of the epidemiology studies it reviewed (studies were judged to be strong, moderately strong, or weak). *Limited evidence* was defined by the committee as evidence from two or more strong or moderately strong studies with varied study designs and populations that suggested an association between exposure to formaldehyde and a specific cancer type, but alternative explanations, such as chance, bias, or confounding factors, could not be adequately ruled out because of limitations in the studies, and so a causal interpretation could not be accepted with confidence. *Sufficient evidence* was defined by the committee as consistent evidence from two or more strong or moderately strong studies with varied study designs and populations that found an association between exposure to formaldehyde and a specific cancer type and for which chance, bias, and confounding factors could be ruled out with reasonable confidence because of the study methodologies and the strength of the findings.

REVIEW OF THE FORMALDEHYDE PROFILE IN THE NATIONAL TOXICOLOGY PROGRAM 12TH REPORT ON CARCINOGENS

To address the first part of its statement of task, this committee reviewed the formaldehyde substance profile in the NTP's 12th RoC. The committee examined the primary literature cited in NTP's background document for formaldehyde and other literature published by June 10, 2011 (the date when the 12th RoC was released). The headings and structure of the committee's review parallel the major headings that NTP used in the substance profile for formaldehyde. As part of its review, the committee determined whether NTP had described and conducted its literature search appropriately, whether the relevant literature identified during the literature search was cited and sufficiently described in the background document, whether NTP had selected the most informative studies in making its listing determination, and whether NTP's arguments supported its conclusion that formaldehyde is known to be a human carcinogen.

Cancer Studies in Humans

The committee reviewed the "Cancer Studies in Humans" section in the NTP substance profile and the corresponding sections in the background document for formaldehyde. The committee concluded that NTP did a thorough job of describing the epidemiology literature in the background document and synthesizing information about key studies in the substance profile. The committee agrees with NTP's focus on three principal types of cohort and case–control studies in humans: studies of industrial workers, studies of professional groups that have high exposure (embalmers), and studies of general-population cohorts and case–control studies.

On the basis of the committee's definition of limited and sufficient evidence discussed above and its peer review of the substance profile for formaldehyde, it concurs with NTP that there is sufficient evidence in studies that had adequate characterization of relevant exposure metrics to enable a conclusion about human cancer after exposure to formaldehyde. Discussions of chance, bias, confounding factors, and other limitations of the most informative studies in the substance profile are clear and thorough.

Epidemiologic evidence was strongest for an association between formaldehyde exposure and cancers of the nasopharyngeal region and sinonasal cavities and myeloid leukemia. NTP considered the most informative study for evaluating nasopharyngeal cancer to be a case-control study[5] that drew incident

---

[5]Vaughan, T.L., P.A. Stewart, K. Teschke, C.F. Lynch, G.M. Swanson, J.L. Lyon, and M. Berwick. 2000. Occupational exposure to formaldehyde and wood dust and nasopharyngeal carcinoma. Occup. Environ. Med. 57(6):376-384.

cases from five US cancer registries that participated in the Surveillance Epidemiology and End Results program of the National Cancer Institute (NCI). Important corroborating evidence for an association was provided by an NCI industrial worker cohort[6] and several case-control studies.[7] NTP considered the most informative study for evaluating sinonasal cancer to be a pooled analysis[8] of several high-quality case–control studies that shared the same method of exposure assessment. Earlier case–control studies[9] combined as a group provided consistent supporting evidence of an association. The potential confounding of the formaldehyde–sinonasal-cancer association by wood-dust exposure was adequately considered by NTP. NTP considered the most informative studies for evaluating lymphohematopoietic cancers, specifically myeloid leukemia, to be the NCI cohort study of industrial workers exposed to formaldehyde,[10] the

---

[6]Hauptmann, M., J.H. Lubin, P.A. Stewart, R.B. Hayes, and A. Blair. 2004. Mortality from solid cancers among workers in formaldehyde industries. Am. J. Epidemiol. 159(12):1117-1130.

[7]Roush, G.C., J. Walrath, L.T. Stayner, S.A. Kaplan, J.T. Flannery, and A. Blair. 1987. Nasopharyngeal cancer, sinonasal cancer, and occupations related to formaldehyde: A case-control study. J. Natl. Cancer Inst. 79(6):1221-1224; West, S., A. Hildesheim, and M. Dosemerci. 1993. Non-viral risk factors for nasopharyngeal carcinoma in the Philippines: Results from a case-control study. Int. J. Cancer 55(5):722-727; Hildesheim, A., M. Dosemeci, C.C. Chan, C.J. Chen, Y.J. Cheng, M.M. Hsu, I.H. Chen, B.F. Mittl, B. Sun, P.H. Levine, J.Y. Chen, L.A. Brinton, and C.S. Yang. 2001. Occupational exposure to wood, formaldehyde, and solvents and risk of nasopharyngeal carcinoma. Cancer Epidemiol. Biomarkers Prev. 10(11):1145-1153.

[8]Luce, D., A. Leclerc, D. Begin, P.A. Demers, M. Gerin, E. Orlowski, M. Kogevinas, S. Belli, I. Bugel, U. Bolm-Audorff, L.A., Brinton, P. Comba, L. Hardell, R.B. Hayes, C. Magnani, E. Merler, S. Preston-Martin, T.L. Vaughan, W. Zheng, and P. Boffetta. 2002. Sinonasal cancer and occupational exposures: A pooled analysis of 12 case-control studies. Cancer Causes Control 13(2):147-157.

[9]Olsen, J.H., S.P. Jensen, M. Hink, K. Faurbo, N.O. Breum, and O.M. Jensen. 1984. Occupational formaldehyde exposure and increased nasal cancer risk in man. Int. J. Cancer 34(5):639-644; Hayes, R.B., J.W. Raatgever, A. de Bruyn, and M. Gerin. 1986. Cancer of the nasal cavity and paranasal sinuses, and formaldehyde exposure. Int. J. Cancer 37(4):487-492; Olsen, J.H., and S. Asnaes. 1986. Formaldehyde and the risk of squamous cell carcinoma of the sinonasal cavities. Br. J. Ind. Med. 43(11):769-774; Roush, G.C., J. Walrath, L.T. Stayner, S.A. Kaplan, J.T. Flannery, and A. Blair. 1987. Nasopharyngeal cancer, sinonasal cancer, and occupations related to formaldehyde: A case-control study. J. Natl. Cancer Inst. 79(6):1221-1224; Luce, D., M. Gerin, A. Leclerc, J.F. Morcet, J. Brugere, and M. Goldberg. 1993. Sinonasal cancer and occupational exposure to formaldehyde and other substances. Int. J. Cancer. 53(2):224-231.

[10]Beane Freeman, L.E., A. Blair, J.H. Lubin, P.A. Stewart, R.B. Hayes, R.N. Hoover, and M. Hauptmann. 2009. Mortality from lymphohematopoietic malignancies among workers in formaldehyde industries: The National Cancer Institute Cohort. J. Natl. Cancer Inst. 101(10):751-761.

NIOSH garment workers cohort,[11] the cohort of chemical workers in six British factories,[12] and the NCI nested case–control study of embalmers.[13] The committee agrees with NTP's assessment that the evidence for cancer at other sites is insufficient at this time and does not rise to the level of limited evidence of a carcinogenic effect in humans.

Cancer in Experimental Animals

The section "Cancer Studies in Experimental Animals" in NTP's substance profile for formaldehyde discusses the degree of certainty of the carcinogenicity of formaldehyde on the basis of evidence from experimental animal studies. NTP concluded that there is sufficient evidence of carcinogenicity of formaldehyde from experimental animal studies. In NTP's discussion of the specific animal findings, it demonstrated that two components of the RoC listing criteria were met. One component was met because there is evidence in animals of increased incidence of malignant tumors or of a combination of malignant and benign tumors in multiple species (several studies in rats and mice) and multiple tissue types (malignancies of the nasal epithelium and gastrointestinal tract). A second component was met because there is evidence in animals of increased incidence of malignant tumors or of a combination of malignant and benign tumors after exposure by multiple routes (inhalation and oral routes). The committee agrees with NTP's overall conclusions.

Studies on Mechanisms of Carcinogenesis

The section "Studies on Mechanisms of Carcinogenesis" in the substance profile for formaldehyde and the associated sections in the background document describe the scientific evidence and mechanistic knowledge available concerning the carcinogenicity of formaldehyde. NTP focused on the mechanisms related to specific clinical sites of cancers, specifically, nasopharyngeal, sinonasal, and lymphohematopoietic cancers. The committee finds that delineation of

---

[11]Pinkerton, L.E., M.J. Hein, and L.T. Stayner. 2004. Mortality among a cohort of garment workers exposed to formaldehyde: An update. Occup. Environ. Med. 61(3):193-200.

[12]Coggon, D., E.C. Harris, J. Poole, and K.T. Palmer. 2003. Extended follow-up of a cohort of British chemical workers exposed to formaldehyde. J. Natl. Cancer Inst. 95(21):1608-1615.

[13]Hauptmann, M., P.A. Stewart, J.H. Lubin, L.E. Beane Freeman, R.W. Hornung, R.F. Herrick, R.N Hoover, J.F. Fraumeni Jr., A. Blair, and R.B. Hayes. 2009. Mortality from lymphohematopoietic malignancies and brain cancer among embalmers exposed to formaldehyde. J. Natl. Cancer Inst. 101(24):1696-1708.

the available mechanistic evidence into portal-of-entry or systemic effects[14] would have made the background document and the substance profile stronger because the mechanisms of carcinogenicity of highly reactive chemicals, including formaldehyde, can differ between portal-of-entry sites and distal sites that their native forms or metabolites might not readily reach. However, the committee found that such changes of presentation would not affect NTP's overall conclusions in the substance profile for formaldehyde.

The committee concludes that NTP correctly stated in the substance profile that "the mechanisms by which formaldehyde causes cancer are not completely understood." There may be several mechanisms of action involved, and the mechanisms proposed by NTP are not mutually exclusive and might be related. Cytotoxicity-induced cellular-proliferation and genotoxicity are two mechanisms that are supported by available evidence in sinonasal and nasopharyngeal regions where inhaled formaldehyde first comes into contact with the mucous layer of the respiratory tract in mammals. Mechanistic evidence of carcinogenicity at distal sites is more uncertain. The substance profile for formaldehyde acknowledges that there is little evidence that formaldehyde or its metabolites would reach systemic circulation or tissues other than those in direct contact with the agent. Given the uncertainties in the scientific understanding of the potential mechanisms of the systemic effects of formaldehyde, the committee finds that NTP could have explicitly acknowledged, as stated in a previous expert panel's report, that "while it would be desirable to have an accepted mechanism that fully explains the association between formaldehyde exposure and distal cancers, the lack of such mechanism should not detract from the strength of the epidemiological evidence that formaldehyde causes myeloid leukemia."[15]

Summary and Conclusions for the Committee's Review of the Formaldehyde Profile in the National Toxicology Program 12th Report on Carcinogens

The committee found that NTP's background document for formaldehyde describes the strengths and weaknesses of relevant studies in a way that is consistent and balanced. The substance profile appropriately cites studies showing positive associations that support the listing. However, the substance profile

---

[14]*Portal-of-entry* effects are effects that arise from direct interaction of inhaled or ingested formaldehyde with the affected cells or tissues. *Systemic* effects are effects that occur beyond tissues or cells at the portal of entry.

[15]McMartin, K.E., F. Akbar-Khanzadeh, G.A. Boorman, A. DeRoos, P. Demers, L. Peterson, S.M. Rappaport, D.B. Richardson, W.T. Sanderson, and M.S. Sandy. 2009. Part B – Recommendation for Listing Status for Formaldehyde and Scientific Justification for the Recommendation. Formaldehyde Expert Panel Report [online]. Available: http://ntp.niehs.nih.gov/ntp/roc/twelfth/2009/November/FA_PartB.pdf.

would be more complete if it included more discussion on why weaker, uninformative, inconsistent, or conflicting evidence did not alter NTP's conclusions. Although the committee identified that as a limitation in the substance profile, it would not likely alter NTP's final conclusions as presented in the substance profile for formaldehyde.

The committee concludes that NTP comprehensively considered available evidence and applied the listing criteria appropriately in reaching its conclusion. The 12th RoC states that "formaldehyde is known to be a human carcinogen based on sufficient evidence of carcinogenicity from studies in humans and supporting data on mechanisms of carcinogenesis." The committee agrees with NTP's conclusion, which is based on evidence published by June 10, 2011, that formaldehyde is a known human carcinogen.

INDEPENDENT ASSESSMENT OF FORMALDEHYDE

The second part of the committee's task was to conduct an independent assessment of formaldehyde. The committee started with the review it undertook in the first part of its task and the background document that supports the formaldehyde profile in the 12th RoC. It searched for additional peer-reviewed literature that has been published by November 8, 2013, and incorporated relevant human, experimental animal, and mechanistic studies into the independent assessment. The cut-off date for the literature search was chosen to allow the committee time to review the literature within the time constraints of the project schedule. Details of the committee's search strategy, exclusion criteria, and corresponding literature trees are provided in Appendix D of this report. The committee focused its attention on literature that contained primary data, but it also examined published review articles and reviews by other authoritative bodies to ensure that all plausible interpretations of primary data were considered. The committee considered comments presented to it during its first meeting, comments and documents received from other sources during the study process, and independent literature searches carried out by National Research Council staff.

The RoC listing criteria places an emphasis on evidence of carcinogenicity in animals or humans for a listing of "reasonably anticipated to be a human carcinogen", and it places an emphasis specifically on evidence in human studies for a listing of "known to be a human carcinogen". For that reason, the committee's independent assessment includes a detailed discussion of its approach for evaluating the epidemiology literature.

The committee's judgment about the strength of a study depended on both the epidemiologic design elements and the exposure assessment dimensions. Particular attention was paid to the choice of summary measures of exposure. Ideally, an epidemiologist chooses the appropriate measure to summarize exposure data on the basis of an understanding or hypothesis about the pharmacokinetics and pharmacodynamics of the exposure-to-dose and dose-to-response processes. The investigators studying the association between formaldehyde and cancer have little information on which to base that choice. In practice, therefore, it is common and appropriate to test the associations by using several different summary measures, including cumulative exposure, average exposure, duration of exposure, and peak exposure. It is expected that, on average, choosing the wrong metric will result in an underestimation of an association if one exists—that is, it is not expected that choosing the wrong summary measure of exposure will falsely create evidence of an association where one does not exist except by chance.

Another factor that complicates the assessment of risks by alternative metrics is the imprecision and other limitations of the exposure-intensity data on which the summary measures are based. Those data are often only approximations and are likely to have substantial uncertainty. That makes it even more difficult to assert with confidence that one summary measure is more likely than another to be "correct". For those reasons, the committee looked at the measures of association between cancer risk and all the available summary measures presented in each study rather than choosing or preferring one a priori. Furthermore, patterns in disease associations and associated confidence intervals from smaller studies that did not reach traditional significance (that is, a p value less than 0.05 and the exclusion of 1.0 from the 95% CI) were not discarded in the committee's evaluation of the literature; they were weighed as weaker but still relevant evidence of consistency in the results.

The committee reviewed the available literature on the topic of which exposure metrics are more appropriate for environmental and occupational cancer studies. There is a long history of using cumulative exposure (the product of average intensity and exposure duration) as the summary measure of exposure. Cumulative exposure tends to be proportional to disease risk and loss of function for nonmalignant respiratory diseases caused by dusts, such as coal dust, silica, and asbestos. Possibly because of that consistency, cumulative exposure has often been used as the summary measure of exposure for other exposures and other diseases, including cancer. But in the few cases in which data are adequate to examine the relative performance of different exposure metrics, it has been found that cumulative exposure is generally not proportional to cancer risk and should not necessarily be assumed to be the correct summary measure of exposure for cancer risk. Evidence of this finding first came from studies of smoking and lung cancer,[16] asbestos exposure and risk of mesothelioma,[17] both asbestos and silica and risk of lung cancer,[18] and leukemia risk and benzene exposure.[19]

---

[16]Doll, R. and R. Peto. 1978. Cigarette smoking and brochial carcinoma: dose and time relationships among regular smokes and lifelong non-smokers. J. Epideol. Community Health. 32(4):303-313.

[17]Peto, J., H. Seidman, I.J. Selikoff. 1982. Mesothelioma mortality in asbestos workers: implications for models of carcinogenesis and risk assessment. Br. J. Cancer 45(1):124-135.

[18]Zeka, A. 2011. The two-stage clonal expansion model in occupational cancer epidemiology: results from three cohort studies. Occ. Env. Med. 68:618-624.

Although it is unclear whether those examples apply to formaldehyde or whether formaldehyde's carcinogenic effects on nasal or bone marrow cells would be expected to show similar exposure–response dynamics, the committee concluded that there was no compelling reason to prefer findings for one of the standard exposure metrics mentioned above over another.

Summary of Evidence for the Committee's Independent Assessment

The statement of task specifically asked the committee to "integrate the level-of-evidence conclusions, and considering all relevant information in accordance with the RoC listing criteria, make an independent listing recommendation for formaldehyde and provide scientific justification for its recommendation" (Appendix B). The committee notes that the term *integrate* does not have a standard definition in the context of hazard assessment. The committee understood the term in its conventional sense of bringing together parts into a whole. To be listed as "reasonably anticipated as a human carcinogen" or "known to be a human carcinogen", the RoC listing criteria only requires information to be integrated across human studies or across animal studies, and supporting information can be derived from mechanistic studies. Mechanistic information "can be useful for evaluating whether a relevant cancer mechanism is operating in people",[20] but a known mechanism is not required for a substance to be listed in the RoC. In the subsections below, the committee summarizes human, experimental-animal, and mechanistic information on nasopharyngeal and sinonasal cancer and myeloid leukemia. Summaries were not presented for other kinds of cancer because of a lack of strong evidence that formaldehyde exposure causes other types of cancer in humans.

Nasopharyngeal and Sinonasal Cancers

The committee found clear and convincing epidemiologic evidence of an association between formaldehyde exposure and nasopharyngeal cancer and sinonasal cancer in humans. On the basis of evidence of an association between nasopharyngeal cancer and exposure to formaldehyde in two strong studies—a

---

[19]Richardson, D.B., C. Terschuren, and W. Hoffmann. 2008. Occupational risk factors for non-Hodgkin's lymphoma: A population-based case-control study in Northern Germany. Am. J. Ind. Med. 51(4):258-268.
[20]NTP (National Toxicology Program). 2010. Report on Carcinogens Background Document for Formaldehyde, January 22, 2010. U.S. Department of Health and Human Services, Public Health Service, National Toxicology Program, Research Triangle Park, NC [online]. Available: http://ntp.niehs.nih.gov/ntp/roc/twelfth/2009/November/Formaldehyde_BD_Final.pdf.

large case-control study[21] and a large cohort study[22]—and other supporting studies that were judged to be moderately strong,[23] the committee concludes that the relationship is causal and chance, bias, and confounding factors can be ruled out with reasonable confidence. For sinonasal cancer, there is evidence of an association based on a strong, well-conducted pooled case–control study (which used pooled data from 12 separate case–control studies)[24] and other, corroborating studies that were judged to be moderately strong.[25] The committee concludes that the relationship between formaldehyde and sinonasal cancer is causal and

---

[21]Vaughan, T.L., P.A. Stewart, K. Teschke, C.F. Lynch, G.M. Swanson, J.L. Lyon, and M. Berwick. 2000. Occupational exposure to formaldehyde and wood dust and nasopharyngeal carcinoma. Occup. Environ. Med. 57(6):376-384.

[22]Beane Freeman, L.E., A. Blair, J.H. Lubin, P.A. Stewart, R.B. Hayes, R.N. Hoover, and M. Hauptmann. 2013. Mortality from solid tumors among workers in formaldehyde industries: An update of the NCI cohort. Am. J. Ind. Med. 56(9):1015-1026.

[23]Vaughan, T.L., C. Strader, S. Davis, and J.R. Daling. 1986a. Formaldehyde and cancers of the pharynx, sinus and nasal cavity: I. Occupational exposures. Int. J. Cancer 38(5):677-683; Vaughan, T.L., C. Strader, S. Davis, and J.R. Daling. 1986b. Formaldehyde and cancers of the pharynx, sinus and nasal cavity: II. Residential exposures. Int. J. Cancer 38(5):685-688; West, S., A. Hildesheim, and M. Dosemerci. 1993. Non-viral risk factors for nasopharyngeal carcinoma in the Philippines: Results from a case-control study. Int. J. Cancer 55(5):722-727; Hildesheim, A., M. Dosemeci, C.C. Chan, C.J. Chen, Y.J. Cheng, M.M. Hsu, I.H. Chen, B.F. Mittl, B. Sun, P.H. Levine, J.Y. Chen, L.A. Brinton, and C.S. Yang. 2001. Occupational exposure to wood, formaldehyde, and solvents and risk of nasopharyngeal carcinoma. Cancer Epidemiol. Biomarkers Prev. 10(11):1145-1153; Siew, S.S., T. Kauppinen, P. Kyyronen, P. Heikkila, and E. Pukkala. 2012. Occupational exposure to wood dust and formaldehyde and risk of nasal, nasopharyngeal, and lung cancer among Finnish men. Cancer. Manag. Res. 4:223-232.

[24]Luce, D., A. Leclerc, D. Begin, P.A. Demers, M. Gerin, E. Orlowski, M. Kogevinas, S. Belli, I. Bugel, U. Bolm-Audorff, L.A., Brinton, P. Comba, L. Hardell, R.B. Hayes, C. Magnani, E. Merler, S. Preston-Martin, T.L. Vaughan, W. Zheng, and P. Boffetta. 2002. Sinonasal cancer and occupational exposures: A pooled analysis of 12 case-control studies. Cancer Causes Control 13(2):147-157.

[25]Hayes, R.B., J.W. Raatgever, A. de Bruyn, and M. Gerin. 1986. Cancer of the nasal cavity and paranasal sinuses, and formaldehyde exposure. Int. J. Cancer 37(4):487-492; Olsen, J.H., and S. Asnaes. 1986. Formaldehyde and the risk of squamous cell carcinoma of the sinonasal cavities. Br. J. Ind. Med. 43(11):769-774; Vaughan, T.L., C. Strader, S. Davis, and J.R. Daling. 1986a. Formaldehyde and cancers of the pharynx, sinus and nasal cavity: I. Occupational exposures. Int. J. Cancer 38(5):677-683; Vaughan, T.L., C. Strader, S. Davis, and J.R. Daling. 1986b. Formaldehyde and cancers of the pharynx, sinus and nasal cavity: II. Residential exposures. Int. J. Cancer 38(5):685-688; Luce, D., M. Gerin, A. Leclerc, J.F. Morcet, J. Brugere, and M. Goldberg. 1993. Sinonasal cancer and occupational exposure to formaldehyde and other substances. Int. J. Cancer. 53(2):224-231; Siew, S.S., T. Kauppinen, P. Kyyronen, P. Heikkila, and E. Pukkala. 2012. Occupational exposure to wood dust and formaldehyde and risk of nasal, nasopharyngeal, and lung cancer among Finnish men. Cancer. Manag. Res. 4:223-232.

Summary

chance, bias, and confounding factors can be ruled out with reasonable confidence.

Several well-conducted studies in experimental animal models demonstrate an increase in nasal squamous cell-carcinoma after inhalation exposure to formaldehyde. Two of the studies used F344 rats,[26] and one used Sprague Dawley rats.[27] The evidence is corroborated by other rat studies[28] and by a study in mice.[29] Although there are limitations in extrapolating findings on nasal tumors in rodents to nasopharyngeal and sinonasal cancers in humans, the experimental-animal evidence indicates that exposure to inhaled formaldehyde is associated with carcinogenic effects on tissues at the portal of entry.

Inhalation of formaldehyde at sufficient concentrations substantially increases formaldehyde to above the total endogenous concentration in tissues at the portal of entry in both animal and human studies. There is experimental evidence that, due to its chemical reactivity, formaldehyde exerts genotoxic and mutagenic effects and cytotoxicity followed by compensatory cell proliferation at the portal of entry in animals and humans exposed to formaldehyde; this provides biologic plausibility of a relationship between formaldehyde exposure and cancer. The evidence on formaldehyde-associated DNA adducts, DNA–protein cross-links, DNA strand breaks, mutations, micronuclei, and chromosomal aberrations is consistent, strong, and specific. In addition, both temporal and exposure–response relationships have been established, most strongly in studies of rodents and nonhuman primates.

---

[26]Kerns, W.D., K.L. Pavkov, D.J. Donofrio, E.J. Gralla, and J.A. Swenberg. 1983. Carcinogenicity of formaldehyde in rats and mice after long-term inhalation exposure. Cancer Res. 43(9):4382-4392; Monticello, T.M., J.A. Swenberg, E.A. Gross, J.R. Leininger, J.S. Kimbell, S. Seilkop, T.B. Starr, J.E. Gibson, and K.T. Morgan. 1996. Correlation of regional and nonlinear formaldehyde-induced nasal cancer with proliferating populations of cells. Cancer Res. 56(5):1012-1022.

[27]Sellakumar, A.R., C.A. Snyder, J.J. Solomon, and R.E. Albert. 1985. Carcinogenicity of formaldehyde and hydrogen chloride in rats. Toxicol. Appl. Pharmacol. 81(3 Pt 1):401-406.

[28]Feron, V.J., J.P. Bruyntjes, R.A. Woutersen, H.R. Immel, and L.M. Appelman. 1988. Nasal tymours in rats after short-term exposure to a cytotoxic concentration of formaldehyde. Cancer Lett. 39(1):101-111; Soffritti, M., C. Maltoni, F. Maffei, and R. Biagi. 1989. Formaldehyde: An experimental multipotential carcinogen. Toxicol. Ind. Health 5(5):699-730; Woutersen, R.A., A. van Garderen-Hoetmer, J.P. Bruijntjes, A. Zwart, and V.J. Feron. 1989. Nasal tumors in rats after severe injury to the nasal mucosa and prolonged exposure to 10ppm formaldehyde. J. Appl. Toxicol. 9(1):39-46; Kamata, E., M. Nakadate, O. Uchida, Y. Ogawa, S. Suzuki, T. Kaneko, M. Saito, and Y. Jurokawa. 1997. Results of a 28-month chronic inhalation toxicity study of formaldehyde in male Fisher-344 rats. J. Toxicol. Sci. 22(3):239-254.

[29]Kerns, W.D., K.L. Pavkov, D.J. Donofrio, E.J. Gralla, and J.A. Swenberg. 1983. Carcinogenicity of formaldehyde in rats and mice after long-term inhalation exposure. Cancer Res. 43(9):4382-4392.

Myeloid Leukemia

The committee found clear and convincing epidemiologic evidence of an association between formaldehyde exposure and myeloid leukemia. There may also be an increase of other lymphohematopoietic cancers, although the evidence is less robust. On the basis of three strong studies with widely different coexposures (NCI formaldehyde industry cohort,[30] NIOSH garment workers cohort,[31] NCI funeral industry cohort[32]) and several moderately strong studies,[33] the committee concludes that there is a causal association between formaldehyde exposure and myeloid leukemia. Chance, bias, and confounding factors can be ruled out with reasonable confidence given the consistent pattern of association in the larger studies that had good exposure assessment.

Although multiple lines of reasoning and experimental evidence indicate that it is unlikely that inhalation exposure to formaldehyde will increase formaldehyde to substantially above endogenous concentrations in tissues distant from the site of entry, there is a robust database of experimental studies of *systemic* mechanistic events that have been observed after exposure to formaldehyde (as discussed in detail in the section "Mechanisms of Carcinogenesis" of Chapter 3). The committee notes that it is plausible that some of the systemic effects, notably findings of genotoxicity and transcriptional changes in circulating blood cells, may have resulted from the exposure of the cells at the portal of entry (for example, lymphoid tissue in the nasal mucosa). The mechanistic events that were considered by the committee to be relevant to the plausibility of formaldehyde-associated tumors beyond the portal of entry included genotoxicity and mutagenicity, hematologic effects, and effects on gene expression. Overall, in mechanistic studies of experimental animals and exposed humans, the evidence

---

[30]Beane Freeman, L.E., A. Blair, J.H. Lubin, P.A. Stewart, R.B. Hayes, R.N. Hoover, and M. Hauptmann. 2009. Mortality from lymphohematopoietic malignancies among workers in formaldehyde industries: The National Cancer Institute Cohort. J. Natl. Cancer Inst. 101(10):751-761.

[31]Meyers, A.R., L.E. Pinkerton, and M.J. Hein. 2013. Cohort mortality study of garment industry workers exposed to formaldehyde: update and internal comparisons. A. J. Ind. Med. 56:1027-1039.

[32]Hauptmann, M., P.A. Stewart, J.H. Lubin, L.E. Beane Freeman, R.W. Hornung, R.F. Herrick, R.N Hoover, J.F. Fraumeni Jr., A. Blair, and R.B. Hayes. 2009. Mortality from lymphohematopoietic malignancies and brain cancer among embalmers exposed to formaldehyde. J. Natl. Cancer Inst. 101(24):1696-1708.

[33]Walrath, J., and J.F. Fraumeni, Jr. 1983. Mortality patterns among embalmers. Int. J. Cancer 31(4):407-411; Walrath, J., and J.F. Fraumeni, Jr. 1984. Cancer and other causes of death among embalmers. Cancer Res. 44(10):4638-4641; Stroup, N.E., A. Blair, and G.E. Erikson. 1986. Brain cancer and other causes of death in anatomists. J. Natl. Cancer Inst. 77(6):1217-1224; Coggon, D., G. Ntani, E.C. Harris, and K.T. Palmer. 2014. Upper airway cancer, myeloid leukemia, and other cancers in a cohort of British chemical workers exposed to formaldehyde. Am. J. Epidemiol. 179(11):1301-1311.

is largely consistent and strong. Both temporal and exposure–response relationships have been demonstrated in studies of humans and animals exposed to formaldehyde. The committee concludes that these findings provide plausible mechanistic pathways supporting a relationship between formaldehyde exposure and cancer, even though the potential mechanisms of how formaldehyde may cause such systemic effects are not fully understood. It would be desirable to have a more complete understanding about how formaldehyde exposure may cause systemic effects, but the lack of known mechanisms should not detract from the findings of an association between formaldehyde exposure and myeloid leukemia in epidemiology studies.

The animal cancer bioassay literature provided some information relevant to myeloid leukemia. One drinking water study[34] reported a significant increase in lymphohematopoietic cancers following long-term exposure to formaldehyde in drinking water, but there is uncertainty regarding the finding. Of the three inhalation studies that included histopathologic examinations of non–respiratory tract tissues, two did not report leukemia.[35] The full laboratory report[36] of a third study[37] discussed findings of leukemia and lymphoma that were not found to be compound related; however, diffuse multifocal bone marrow hyperplasia was observed in some male and female rats. Although that finding was not a finding of malignancy, it does indicate that long-term inhaled formaldehyde may cause effects in bone marrow.

Final Conclusions and Listing Recommendation

The committee identified and evaluated relevant, publicly available, peer-reviewed literature on formaldehyde, including attention to literature published between June 10, 2011 (the release date of the substance profile for formaldehyde in the 12th RoC), and November 8, 2013. The committee applied NTP's established RoC listing criteria to the scientific evidence on formaldehyde from

---

[34]Soffritti, M. F. Belpoggi, L. Lambertin, M. Lauriola, M. Padovani, and C. Maltoni. 2002. Results of long-term exposreimental studies on the carcinogeneicity of formaldehyde and acetaldehyde in rats. Ann. N.Y. Acad. Sci. 982:87-105.

[35]Sellakumar, A.R., C.A. Snyder, J.J. Solomon, and R.E. Albert. 1985. Carcinogenicity of formaldehyde and hydrogen chloride in rats. Toxicol. Appl. Pharmacol. 81(3 Pt 1):401-406; Kamata, E., M. Nakadate, O. Uchida, Y. Ogawa, S. Suzuki, T. Kaneko, M. Saito, and Y. Jurokawa. 1997. Results of a 28-month chronic inhalation toxicity study of formaldehyde in male Fisher-344 rats. J. Toxicol. Sci. 22(3):239-254.

[36]Battelle. 1981. Final Report on a Chronic Inhalation Toxicology Study in Rats and Mice Exposed to Formaldehyde. Prepared by Battelle Columbus Laboratories, Columbus, OH, for the Chemical Industry Institute of Toxicology (CIIT), Research Triangle Park, NC. CIIT Docket No. 10922.

[37]Kerns, W.D., K.L. Pavkov, D.J. Donofrio, E.J. Gralla, and J.A. Swenberg. 1983. Carcinogenicity of formaldehyde in rats and mice after long-term inhalation exposure. Cancer Res. 43(9):4382-4392.

studies of humans, studies of experimental animals, and other studies relevant to mechanisms of carcinogenesis.

The type of information needed to meet the criteria for sufficient evidence in experimental animals is clear and transparent, as discussed above. In contrast, the RoC listing criteria do not provide detailed guidance about how evidence should be assembled to meet the requirement of limited evidence or sufficient evidence of carcinogenicity from studies in humans, except to note that limited evidence cannot exclude alternative explanations, such as chance, bias, or confounding factors, and to note that conclusions should be based on "scientific judgment, with consideration given to all relevant information".[38] To evaluate the epidemiology evidence, the committee used scientific judgment to develop an approach to assessing the epidemiology evidence. The approach included careful review of individual studies, selection of studies that were most informative, and evaluation of informative studies on the basis of the strength, consistency, temporality, dose-response, and coherence of the evidence.

The committee notes that evidence in experimental animals and a known mechanism of action is not required by the RoC listing criteria in making a listing recommendation that a substance is known to be a human carcinogen if the evidence from studies in humans is sufficient and indicates an association between exposure and human cancer. Also, and importantly, the RoC listing criteria require an association in only one type of cancer to make the determination. On the basis of the information summarized directly above for nasopharyngeal and sinonasal cancers and for myeloid leukemia, the committee makes its independent determinations as follows:

- There is sufficient evidence of carcinogenicity from studies of humans based on consistent epidemiologic findings on nasopharyngeal cancer, sinonasal cancer, and myeloid leukemia for which chance, bias, and confounding factors can be ruled out with reasonable confidence.
- There is sufficient evidence of carcinogenicity in animals based on malignant and benign tumors in multiple species, at multiple sites, by multiple routes of exposure, and to an unusual degree with regard to type of tumor.
- There is convincing relevant information that formaldehyde induces mechanistic events associated with the development of cancer in humans, specifically genotoxicity and mutagenicity, hematologic effects, and effects on gene expression.

---

[38]NTP (National Toxicology Program). 2010. Report on Carcinogens Background Document for Formaldehyde, January 22, 2010. U.S. Department of Health and Human Services, Public Health Service, National Toxicology Program, Research Triangle Park, NC [online]. Available: http://ntp.niehs.nih.gov/ntp/roc/twelfth/2009/November/Formaldehyde_BD_Final.pdf.

Because there is sufficient evidence of carcinogenicity from studies in humans that indicates a causal relationship between exposure to formaldehyde and at least one type of human cancer, the committee concludes that formaldehyde should be listed in the RoC as "known to be a human carcinogen".

1

Introduction

Many people in the United States are exposed to formaldehyde from environmental sources (for example, combustion processes, building materials, and tobacco smoke) or in occupational settings (for example, the furniture, textile, and construction industries) (NTP 2011a; IARC 2012). Scientists have studied formaldehyde for decades to determine whether exogenous formaldehyde exposure might be associated with cancer in humans. Much of the focus has been on cancers of the upper respiratory tract because they were thought to be the most biologically plausible (Collins and Lineker 2004). However, there is increasing interest in a potential relationship between formaldehyde exposure and some lymphohematopoietic cancers (for example, leukemia) (NTP 2010a; IARC 2012).

The National Toxicology Program (NTP) first assessed the potential carcinogenicity of formaldehyde in the late 1970s and early 1980s, and the substance was listed as "reasonably anticipated to be a human carcinogen" in the 2nd Report on Carcinogens (RoC) (NTP 1981). Three decades later, NTP reassessed formaldehyde and upgraded its listing to "known to be a human carcinogen" in the 12th RoC (NTP 2011a). In 2012, Congress directed the Department of Health and Human Services (DHHS) to contract with the National Academy of Sciences to carry out an independent review of the formaldehyde substance profile in the 12th RoC (112th Congress, 1st Session; Public Law 112-74). This report presents findings and conclusions in response to the congressional request.

THE REPORT ON CARCINOGENS

NTP is an interagency program involving the National Institutes of Health's National Institute of Environmental Health Sciences (NIEHS, the administrative lead), the Centers for Disease Control and Prevention's National Institute for Occupational Safety and Health, and the Food and Drug Administration's National Center for Toxicological Research. It currently publishes the RoC, which was congressionally mandated in 1978 as part of the Public Health Service Act (Section 262, Public Law 95-622, Part E). The act directed DHHS to publish an annual report that includes a list of all substances that meet two conditions: a significant number of people living in the United States are exposed and the substance is either known to be a carcinogen or may reasonably be anticipated to be a carcinogen. The report was also required to include supporting information, such as the nature of exposure and an estimated number of persons exposed. The full congressional mandate is in Box 1-1. In 1993, an amendment moved the RoC from an annual to a biennial report (42 US Code 241).

Nominations for substances to be added, reclassified, or removed from the RoC can come from anyone, but the submitter must include a rationale and, if possible, background information to support the addition, reclassification, or removal (NTP 2011b). Staff of the Office of the Report on Carcinogens review each submission and decide whether a substance should move forward for further evaluation. From that point, staff of the office invite partnering agencies to review the substance, solicit public comments through the *Federal Register*, and develop a brief draft concept document with information on the substance, exposure, major relevant issues, and an approach to the cancer-evaluation component of an ROC. After consideration of comments from NTP's Board of Scientific Counselors and public comments, the NTP director makes the final decision as to whether the substance will be evaluated in an RoC.

Each RoC is cumulative and includes substances listed as "known to be a human carcinogen" or "reasonably anticipated to be a human carcinogen" since the 1st RoC in 1980. The 12th RoC contains 240 listings; 54 substances are listed as known human carcinogens and 186 as reasonably anticipated to be human carcinogens. The criteria that are currently used to guide the establishment of a listing as either known or reasonably anticipated to be a human carcinogen have been in use since the 8th RoC, published in 1998. Box 1-2 provides the specific listing criteria.

In preparation for a new RoC, the Office of the Report on Carcinogens creates a background document for each substance, which describes in detail properties, production and use, human exposure, toxicokinetics, cancer studies in humans and animals, and mechanisms of action of cancer induction. The purpose of the background document is to describe the strengths, limitations, and overall quality of studies that make up the scientific body of evidence for or against carcinogenicity. For the 12th RoC, background documents for reclassified or newly listed substances were reviewed by an expert panel, and the panel was asked to recommend a listing status for each substance in accordance with the RoC listing criteria (see Figure 1-1 for a depiction of the 12th RoC process). An Interagency Scientific Review Group and an NIEHS–NTP Scientific Review Group were also asked to review each background document and to recommend a listing status. A corresponding draft substance profile was then prepared by NTP on the basis of the background document, the aforementioned reviews, and the listing recommendations, and the draft profile was reviewed by the NTP Board of Scientific Counselors. Public comments were solicited at multiple stages in the process. At the end of the process, the profiles of all 240 substances were compiled into a draft RoC that was submitted to the NTP director for review; to the NTP Executive Committee[1] for consultation, review, and comment; to the NTP director again for final approval; and finally to the secretary of health and human services for review, approval, and transmittal to Congress. The 12th RoC was published on June 10, 2011.

BOX 1-1 Congressional Language Mandating the Report on Carcinogens

A. a list of all substances
  i. which either are known to be carcinogens or may reasonably be anticipated to be carcinogens and
  ii. to which a significant number of persons residing in the United States are exposed;
B. information concerning the nature of such exposure and the estimated number of persons exposed to such substances;
C. a statement identifying
  i. each substance contained in the list under subparagraph (A) for which no effluent, ambient, or exposure standard has been established by a Federal agency, and
  ii. for each effluent, ambient, or exposure standard established by a Federal agency with respect to a substance contained in the list under subparagraph (A), the extent to which, on the basis of available medical, scientific, or other data, such standard, and the implementation of such standard by the agency, decreases the risk to public health from exposure to the substance; and
D. a description of
  i. each request received during the year involved
    I. from a Federal agency outside the Department of Health, Education, and Welfare for the Secretary, or
    II. from an entity within the Department of Health, Education, and Welfare to any other entity within the Department, to conduct research into, or testing for, the carcinogenicity of substances or to provide information described in clause (ii) of subparagraph (C), and
  ii. how the Secretary and each such other entity, respectively, have responded to each such request.

Source: Section 262, Public Law 95-622, Part E (pp. 3435-3436).

---

[1] The NTP Executive Committee is made up of the heads of the Consumer Product Safety Commission, the Department of Defense, the Environmental Protection Agency, the Food and Drug Administration, the National Cancer Institute, the National Center for Environmental Health, the Agency for Toxic Substances and Disease Registry, the National Institute of Environmental Health Sciences, the National Institute for Occupational Safety and Health, and the Occupational Safety and Health Administration. The committee gives programmatic and policy advice to the NTP director.

> **BOX 1-2** Listing Criteria for the Report on Carcinogens
>
> ***Known To Be Human Carcinogen:***
>
> There is sufficient evidence of carcinogenicity from studies in humans,* which indicates a causal relationship between exposure to the agent, substance, or mixture, and human cancer.
>
> ***Reasonably Anticipated To Be Human Carcinogen:***
>
> There is limited evidence of carcinogenicity from studies in humans,* which indicates that causal interpretation is credible, but that alternative explanations, such as chance, bias, or confounding factors, could not adequately be excluded,
>
> or
>
> there is sufficient evidence of carcinogenicity from studies in experimental animals, which indicates there is an increased incidence of malignant and/or a combination of malignant and benign tumors (1) in multiple species or at multiple tissue sites, or (2) by multiple routes of exposure, or (3) to an unusual degree with regard to incidence, site, or type of tumor, or age at onset,
>
> or
>
> there is less than sufficient evidence of carcinogenicity in humans or laboratory animals; however, the agent, substance, or mixture belongs to a well-defined, structurally related class of substances whose members are listed in a previous Report on Carcinogens as either known to be a human carcinogen or reasonably anticipated to be a human carcinogen, or there is convincing relevant information that the agent acts through mechanisms indicating it would likely cause cancer in humans.
>
> Conclusions regarding carcinogenicity in humans or experimental animals are based on scientific judgment, with consideration given to all relevant information. Relevant information includes, but is not limited to, dose response, route of exposure, chemical structure, metabolism, pharmacokinetics, sensitive sub-populations, genetic effects, or other data relating to mechanism of action or factors that may be unique to a given substance. For example, there may be substances for which there is evidence of carcinogenicity in laboratory animals, but there are compelling data indicating that the agent acts through mechanisms which do not operate in humans and would therefore not reasonably be anticipated to cause cancer in humans.
>
> *This evidence can include traditional cancer epidemiology studies, data from clinical studies, and/or data derived from the study of tissues or cells from humans exposed to the substance in question, which can be useful for evaluating whether a relevant cancer mechanism is operating in humans.
>
> Source: NTP 2010a, p. iv.

24

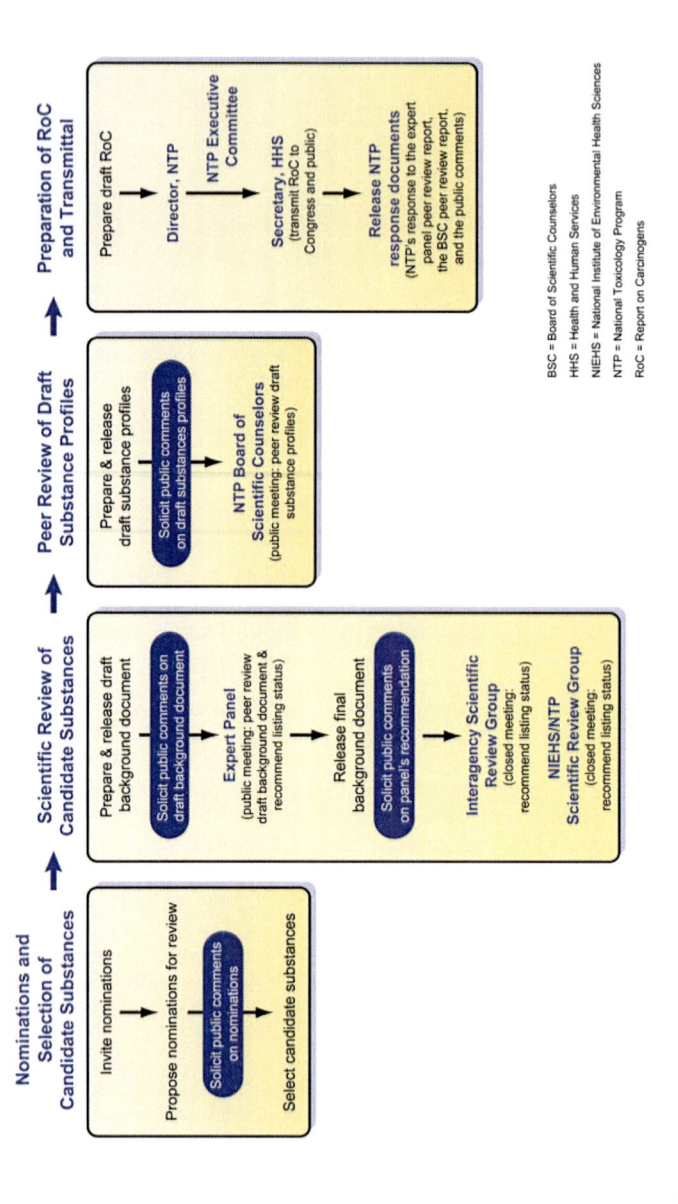

FIGURE 1-1 Schematic of the review process for the 12th Report on Carcinogens. Source: NTP 2011b.

Introduction 25

FORMALDEHYDE AND THE REPORT ON CARCINOGENS

One substance profile in the 12th RoC that has drawn science, policy, and news-media attention is that of formaldehyde (Risk Policy Report 2011a,b; Kristof 2012). Formaldehyde is a colorless gas at room temperature with a pungent smell, has a simple chemical structure, and is one of the most reactive aldehydes (NTP 2010a). It is an economically important chemical in the United States—ranking 25th in overall chemical production—and products that contain formaldehyde account for more than 5% of the annual US gross domestic product (Zhang et al. 2009). The most common use of formaldehyde is in the production of synthetic resins, such as urea– and phenol–formaldehyde resins, that are used as adhesives in particleboard, fiberboard, and plywood. Formaldehyde is also used in textiles to make materials creaseproof, crushproof, flame-resistant, and shrinkproof; to mold plastic parts for automobiles, home appliances, hardware, garden equipment, and sporting equipment; to preserve dried food, fish, oils, and fats; to disinfect containers in the food industry; and, in agriculture, as a preservative, fumigant, germicide, fungicide, and insecticide. In a smaller market, formaldehyde is used in medicines to modify and reduce the toxicity of viruses, venoms, and irritating pollens (ATSDR 1999; NTP 2010a).

Characterizing exposure to formaldehyde and linking exposure to disease are complicated by the many possible sources of exposure, both environmental and occupational. Epidemiologic studies undertaken to understand the potential linkage are sometimes confounded by exposures to other agents known to cause disease, such as cigarette smoke or wood-dust particles. An additional complexity is the fact that formaldehyde is produced naturally in humans and other animals (IARC 2006; NTP 2010a). The chemical "is an essential metabolic intermediate in all cells and is produced endogenously from serine, glycine, ethionine, and choline, and from the demethylation of N-, O-, S-, methyl compounds" (NTP 2010a, p.14).

Formaldehyde was first listed in the 2nd RoC (1981) as reasonably anticipated to be a human carcinogen. However, it was nominated for possible reclassification by NIEHS on the basis of the International Agency for Research on Cancer (IARC) review of the substance in 2004 (NTP 2007). IARC has reviewed formaldehyde several times, concluding with increasing certainty that formaldehyde causes cancer in humans. In 1982, it was classified as "possibly carcinogenic to humans" (IARC 1982); in 1987 and 1995, it was classified as "probably carcinogenic to humans" (IARC 1987, 1995); and in 2006, IARC "concluded that there was sufficient evidence for the carcinogenicity of formaldehyde in humans" (IARC 2006). IARC again listed formaldehyde as carcinogenic to humans in another recent review (IARC 2012).

Formaldehyde was accepted by NTP for review and possible reclassification, and it was reviewed according to established NTP policies and procedures. NTP released a final background document for the assessment of formaldehyde in January 2010, and the substance profile for formaldehyde was published in June 2011 as part of the 12th RoC. In the 12th RoC, formaldehyde was listed as known to be a human carcinogen on the basis of the listing criteria described in Box 1-2 and the supporting information provided in the background document (NTP 2010a, 2011a).

THE COMMITTEE'S TASK

Congress directed DHHS to arrange for the National Academy of Sciences to conduct an independent scientific peer review of the 12th Report on Carcinogens determinations related to formaldehyde and styrene. The request was made in 2012 as part of the Consolidated Appropriations Act (112th Congress, 1st Session; Public Law 112-74). In response, the National Research Council convened the Committee to Review the Formaldehyde Assessment in the National Toxicology Program 12th Report on Carcinogens, which wrote the present report. The committee included experts in epidemiology, exposure assessment, toxicology, toxicokinetic modeling, and mechanisms of carcinogenesis (see Appendix A for biographic information on the committee).

The committee's Statement of Task is presented in Appendix B. The committee was asked to conduct a peer review of the formaldehyde assessment in the 12th RoC. As part of that review, it was asked to identify and evaluate relevant peer-reviewed scientific literature, with emphasis on literature that had been published by June 10, 2011, the release date of the 12th RoC. The committee was also asked to undertake an independent assessment of formaldehyde, which was to include documentation of its decisions related to inclusion or exclusion of literature, identification of critical studies and information, application of the RoC listing criteria to the scientific evidence, and making independent level-of-evidence determinations with respect to the human and animal studies. Considering all relevant information in accordance with the RoC listing criteria, the committee was asked to make an independent listing recommendation for formaldehyde and provide scientific justification for the recommendation. The committee's listing recommendation is based on "scientific judgment, with consideration given to all relevant information", as instructed in the RoC listing criteria.

THE COMMITTEE'S APPROACH

In writing its report, the committee reviewed documents pertaining to formaldehyde that were written for or by NTP in preparation for the 12th RoC (see Table 1-1). It considered presentations heard during its open-session meeting, comments submitted from the general public,[2] and abstracts presented during recent conferences. It reviewed reports published by other authoritative bodies, and it examined primary literature, reviews, and meta-analyses that were

---

[2] A list and copies of materials submitted from the general public can be obtained by contacting the National Academies Public Access Records Office.

TABLE 1-1 Documents Pertaining to Formaldehyde That Were Available to or Written by NTP

Document	Brief Description	Reference
Substance profile for formaldehyde	The substance profile as presented in the 12th RoC	NTP 2011a
Background document for formaldehyde	Background information that was prepared by staff in the Office of the RoC to support NTP's assessment of formaldehyde	NTP 2010a
Primary literature	Primary literature cited in the background document or obtained from other sources	—
Expert panel reports	Reports of an expert panel charged with doing a peer review of the draft background document on formaldehyde and making a recommendation for listing status in the 12th RoC	McMartin et al. 2009, 2010
NTP Executive Committee Interagency Scientific Review Group (ISRG) Report	The interagency scientific review group that reviewed the body of literature on formaldehyde and made a recommendation for the listing of formaldehyde in the 12th RoC	NTP 2010b
National Institute of Environmental Health Sciences (NIEHS)–NTP Scientific Review Group Report	The NIEHS–NTP scientific review group that reviewed the body of literature on formaldehyde and made a recommendation for the listing of formaldehyde in the 12th RoC	NTP 2010c
Minutes from the Board of Scientific Counselors (BSC) Meeting	Report of BSC's assessment of whether the scientific information in the draft substance profile is technically correct, is clearly stated, and supports NTP's preliminary listing of formaldehyde in the 12th RoC	NTP 2010d
NTP's response to the expert panel reports and to BSC	NTP's review of and response to expert panel reports	NTP 2011c, 2011d
Public comments	Comments from the public in response to *Federal Register* notices on October 18, 2005 (Vol. 70, No. 200), August 31, 2009 (Vol. 74, No. 167), December 21, 2009 (Vol. 74, No. 243), and April 22, 2010 (Vol. 75, No. 77) and additional public comments that were not associated with any *Federal Register* notices	NTP 2011e
NTP's response to public comments	NTP's responses to public comments related to specific issues in the expert panel reports that were applicable to the substance profile (comments on the final background document, the review process, or nontechnical or nonscientific issues were excluded by NTP)	NTP 2011f

publicly available in the peer-reviewed literature. The committee was guided by the language and terminology of the RoC listing criteria (see Box 1-2), and it used its own professional judgment for the interpretation of such terms as *sufficient* and *limited*. The committee worked toward the goal of clearly describing its own methods in writing this report, how it used the language of the listing criteria, and its analysis of the body of evidence related to formaldehyde.

The committee noted that the assessment of chemicals for the purposes of listing in the RoC constitutes a hazard assessment, not a risk assessment. A hazard assessment focuses on the identification of substances that may pose a hazard to human health, and it "makes a classification regarding toxicity, for example, whether a chemical is 'carcinogenic to humans' or 'likely to be' (EPA 2005)" (NRC 2009, p. 113). A risk assessment[3] focuses on the likely degree of damage and requires much more information, including completion of a hazard identification, dose–response analysis, exposure quantification, and characterization of risk (NRC 1983). The committee thus approached its assessment of formaldehyde as an evaluation of hazard, not risk. It evaluated measures of association in a population (such as risk ratios, odds ratios, and incidence ratios) from epidemiology studies to inform its assessment of formaldehyde, but it did not identify exposure scenarios that could pose cancer risk as part of a full risk assessment.

This Review and the 2011 Draft IRIS Assessment of Formaldehyde

The committee examined the National Research Council report, *Review of the Environmental Protection Agency's Draft IRIS [Integrated Risk Information System] Assessment of Formaldehyde* (NRC 2011). Although the present report and the 2011 report both focused on formaldehyde, the two committees had different statements of task. The Committee to Review EPA's Draft IRIS Assessment of Formaldehyde was asked to "conduct an independent scientific review of [EPA's] draft human health assessment of formaldehyde for [IRIS]." It was also asked to address specific questions related to EPA's inhalation reference concentration (RfC) for noncancer health effects and its risk estimate for carcinogenicity. That committee assessed how well the narrative presented in the draft IRIS assessment supported the IRIS assessment's conclusions regarding health effects. That committee did not conduct its own literature search, review

---

[3]"Risk assessment is the use of the factual base to define the health effects of exposure of individuals or populations to hazardous materials and situations. . . . Risk assessments contain some or all of the following four steps: Hazard identification: The determination of whether a particular chemical is or is not causally linked to particular health effects. Dose–response assessment: The determination of the relation between the magnitude of exposure and the probability of occurrence of the health effects in question. Exposure assessment: The determination of the extent of human exposure before or after application of regulatory controls. Risk characterization: The description of the nature and often the magnitude of human risk, including attendant uncertainty" (NRC 1983, p. 3).

all relevant evidence, systematically formulate its own conclusions regarding causality, or recommend values for the RfC and unit risk. In contrast, the committee that wrote the present report was asked to identify relevant peer-reviewed literature, document its decisions regarding inclusion or exclusion of the literature, apply NTP's RoC listing criteria, and make an independent listing recommendation for formaldehyde (see Appendix B).

The two projects were also different because of inherent differences between EPA's IRIS assessments and NTP's RoC. IRIS assessments are comprehensive human health assessments that evaluate cancer and noncancer end points and include hazard and dose-response assessments that are used to derive toxicity values (that is, reference values and unit risk values), whereas NTP qualitatively weighs evidence of carcinogenicity and compiles lists of substances that it classifies as known human carcinogens or reasonably anticipated human carcinogens to produce the biennial RoC. Because of those differences, the committee cautions readers against making direct comparisons between the two reports.

This Review and Other Ongoing Studies

The committee that wrote this report worked in parallel with the Committee to Review the Styrene Assessment in the National Toxicology Program 12th Report on Carcinogens, which was also convened in response to the 2012 Consolidated Appropriations Act. The two committees' statements of task were identical except for the specific substance profiles being reviewed, and they met jointly for their first meeting. During the open session of that meeting, the committees heard presentations from and had an open discussion with representatives of DHHS and NTP. Several stakeholders also participated in the public session. During the meeting's closed session, members discussed the open-session presentations by the sponsor and the public and the committees' approach to their statements of task. The two committees also discussed general approaches to the domains of evidence to be examined (specifically, epidemiology, experimental animal studies, and mechanistic information). In particular, the committees discussed an approach that considered principles of the Bradford Hill criteria with respect to causality and an approach to make judgments about individual studies and about the overall body of evidence pertaining to exposure to a substance and cancer. No discussions took place between the full committees after that first joint meeting. The membership of the two committees included three overlapping members who ensured that the committees continued to have compatible approaches to their statements of task.

The committee was also cognizant of the ongoing work of the Committee to Review EPA's IRIS Process (NRC 2014). Part of that committee's task was to "review current methods for evidence-based reviews and recommend approaches for weighing scientific evidence for chemical hazard and dose–response assessments" (NRC 2014). Because the Committee to Review EPA's

IRIS Process and the present committee wrote their reports concurrently, the methods of the present report could not be informed by the conclusions and recommendations of the other one. However, the final report of the Committee to Review the IRIS Process goes beyond recommendations that are only applicable to the IRIS process. It includes discussions on best practices for systematically weighing and integrating scientific evidence that could be used to inform listing determinations in future editions of the RoC.

ORGANIZATION OF THE REPORT

The committee approached its statement of task by first conducting a review of the substance profile for formaldehyde as presented in the 12th RoC (Chapter 2). It considered literature published by June 10, 2011. Chapter 2 is organized on the basis of the headings and subheadings of the substance profile and concludes with a listing recommendation for formaldehyde that is based on the application of the RoC listing criteria to the evidence in the background document and substance profile for formaldehyde. The committee then conducted its own independent assessment of the formaldehyde literature (Chapter 3), extending its review to include literature through November 8, 2013, and concluding with its own listing recommendation for formaldehyde. Appendix A presents the biographies of the committee members, and Appendix B reproduces the committee's statement of task. Appendix C discusses exposure assessment for epidemiologic carcinogenicity studies, Appendix D describes the literature search strategies used to support the evidence presented in Chapter 3, and Appendix E contains summary tables to supplement the genotoxicity and mutagenicity section of Chapter 3.

REFERENCES

ATSDR (Agency for Toxic Substances and Disease Registry). 1999. Toxicological Profile for Formaldehyde. U.S. Department of Health and Human Services, Public Health Service, Agency for Toxic Substances and Disease Registry, Atlanta, GA [online]. Available: http://www.atsdr.cdc.gov/toxprofiles/tp111.pdf [accessed Sept. 23, 2013].

Collins, J.J., and G.A. Lineker. 2004. A review and meta-analysis of formaldehyde exposure and leukemia. Regul. Toxicol. Pharmacol. 40(2):81-91.

EPA (Environmental Protection Agency). 2005.Guidelines for Carcinogen Risk Assessment. EPA/630/P-03/001F. Risk Assessment Forum, U.S. Environmental Protection Agency, Washington, DC. March 2005.

IARC (International Agency for Research on Cancer). 1982. Chemicals, Industrial Processes and Industries Associated with Cancer in Humans: An Updating of IARC Monographs Volumes 1 to 29. IARC Monographs on the Evaluation of the Carcinogenic Risks to Humans Supplement 4. Lyon, France: IARC [online]. Available: http://monographs.iarc.fr/ENG/Monographs/suppl4/Suppl4.pdf [accessed June 10, 2013].

IARC (International Agency for Research on Cancer). 1987. Overall Evaluations of Carcinogenicity: An Updating of IARC Monographs Volumes 1 to 42. IARC Monographs on the Evaluation of the Carcinogenic Risks to Humans Supplement 7. Lyon, France: IARC [online]. Available: http://monographs.iarc.fr/ENG/Monographs/suppl7/Suppl7.pdf [accessed June. 10, 2013].

IARC (International Agency for Research on Cancer). 1995. Wood Dust and Formaldehyde. IARC Monographs on the Evaluation of the Carcinogenic Risks to Humans Vol. 62. Lyon, France: IARC [online]. Available: http://monographs.iarc.fr/ENG/Monographs/vol62/mono62.pdf [accessed June 10, 2013].

IARC (International Agency for Research on Cancer). 2006. Formaldehyde, 2-Butoxyethanol and 1-tert-Butoxy-propan-2-ol. IARC Monographs on the Evaluation of the Carcinogenic Risks to Humans Vol. 88. Lyon, France: IARC [online]. Available: http://monographs.iarc.fr/ENG/Monographs/vol88/mono88.pdf [accessed June 10, 2013].

IARC (International Agency for Research on Cancer). 2012. Chemical Agents and Related Occupations: A Review of Human Carcinogens. IARC Monographs on the Evaluation of the Carcinogenic Risks to Humans Vol. 100F. Lyon, France: IARC [online]. Available: http://monographs.iarc.fr/ENG/Monographs/vol100F/mono100F.pdf [accessed June 10, 2013].

Kristof, N.D. 2012. The Cancer Lobby. The New York Times, October 6, 2012 [online]. Available: http://www.nytimes.com/2012/10/07/opinion/sunday/kristof-the-cancer-lobby.html?_r=0 [accessed June 10, 2013].

McMartin, K.E., F. Akbar-Khanzadeh, G.A. Boorman, A. DeRoos, P. Demers, L. Peterson, S.M. Rappaport, D.B. Richardson, W.T. Sanderson, and M.S. Sandy. 2009. Part B – Recommendation for Listing Status for Formaldehyde and Scientific Justification for the Recommendation. Formaldehyde Expert Panel Report [online]. Available: http://ntp.niehs.nih.gov/ntp/roc/twelfth/2009/November/FA_PartB.pdf [accessed July 17, 2013].

McMartin, K.E., F. Akbar-Khanzadeh, G.A. Boorman, A. DeRoos, P. Demers, L. Peterson, S.M. Rappaport, D.B. Richardson, W.T. Sanderson, and M.S. Sandy. 2010. Part A – Peer Review of the Draft Background Document on Formaldehyde. Formaldehyde Expert Panel Report [online]. Available: http://ntp.niehs.nih.gov/ntp/roc/twelfth/2009/november/fa_parta.pdf [accessed July 17, 2013].

NRC (National Research Countil). 1983. Risk Assessment in the Federal Government: Managing the Process. Washington, DC: National Academy Press.

NRC (National Research Council). 2009. Science and Decision: Advancing Risk Assessment. Washington, DC: National Academies Press.

NRC (National Research Council). 2011. Review of the Environmental Protection Agency's Draft IRIS Assessment of Formaldehyde. Washington, DC: National Academies Press.

NRC (National Research Council). 2014. Statement of Task for the Committee to Review the Integrated Risk Information System Process [online]. Available: http://www8.nationalacademies.org/cp/projectview.aspx?key=49458. [accessed Feb. 20, 2014].

NTP (National Toxicology Program). 1981. Second Annual Report on Carcinogens. U.S. Department of Health and Human Services, Public Health Service, National Toxicology Program, Research Triangle Park, NC.

NTP (National Toxicology Program). 2007. Formaldehyde. Nomination Information [online]. Available: http://ntp.niehs.nih.gov/index.cfm?objectid=7BE524E1-F1F6-975E-76BB0ABD6CC9076A [accessed July 18, 2013].

NTP (National Toxicology Program). 2010a. Report on Carcinogens Background Document for Formaldehyde, January 22, 2010. U.S. Department of Health and Human Services, Public Health Service, National Toxicology Program, Research Triangle Park, NC [online]. Available: http://ntp.niehs.nih.gov/ntp/roc/twelfth/2009/November/Formaldehyde_BD_Final.pdf [accessed July 17, 2013].

NTP (National Toxicology Program). 2010b. Recommendation for Listing Status of Formaldehyde. NTP Executive Committee Interagency Scientific Review Group (ISRG). March 17, 2010.

NTP (National Toxicology Program). 2010c. Recommendation for Listing Status for Formaldehyde in the Report on Carcinogens. Report on Carcinogens (RoC) NIEHS/NTP Scientific Review Group (NSRG). March 16, 2010.

NTP (National Toxicology Program). 2010d. Summary Minutes June 21-22, 2010, Board of Scientific Counselors, National Institute of Environmental Health Sciences, National Toxicology Program, Research Triangle Park, NC [online]. Available: http://ntp.niehs.nih.gov/ntp/About_NTP/BSC/2010/June/Minutes20100622.pdf [accessed July 17, 2013].

NTP (National Toxicology Program). 2011a. Formaldehyde. Pp. 195-205 in Report on Carcinogens, 12th Ed. U.S. Department of Health and Human Services, Public Health Service, National Toxicology Program, Research Triangle Park, NC [online]. Available: http://ntp.niehs.nih.gov/ntp/roc/twelfth/profiles/formaldehyde.pdf [accessed July 17, 2013].

NTP (National Toxicology Program). 2011b. Report on Carcinogens, 12th Ed. U.S. Department of Health and Human Services, Public Health Service, National Toxicology Program, Research Triangle Park, NC [online]. Available: http://ntp.niehs.nih.gov/ntp/roc/twelfth/roc12.pdf [accessed July 17, 2013].

NTP (National Toxicology Program). 2011c. NTP Response to Expert Panels' Peer-Review Comments on Background Documents for Candidate Substances for the 12th Report on Carcinogens [online]. Available: http://ntp.niehs.nih.gov/ntp/roc/twelfth/2011/ResponseExpertPanelReport2011.pdf [accessed Oct. 28, 2013].

NTP (National Toxicology Program). 2011d. NTP Response to the NTP Board of Scientific Counselors (BSC) Peer Review Comments on the Draft Substances Profiles for the 12th Report on Carcinogens, June 21-22, 2010, BSC Meeting [online]. Available: http://ntp.niehs.nih.gov/ntp/roc/twelfth/2011/Response062110BSC2011.pdf [accessed Oct. 28, 2013].

NTP (National Toxicology Program). 2011e. Formaldehyde (CAS No. 50-00-0] Public Comments [online]. Available: http://ntp.niehs.nih.gov/?objectid=20A477F2-F1F6-975E-7472FC6B0DA56D9C#formaldehyde [accessed July 17, 2013].

NTP (National Toxicology Program). 2011f. Formaldehyde. Pp. 9-25 in NTP Response to Issues Raised in the Public Comments for Candidate Substances for the 12th Report on Carcinogens[online]. Available: http://ntp.niehs.nih.gov/ntp/roc/twelfth/2011/ResponsePublicComments2011.pdf#page=11 [accessed Oct. 28, 2013].

Risk Policy Report. 2011a. Industry Targets Cancer Report in New Push for Hill Scrutiny of Risk Studies. Risk Policy Report 18(31). August 2, 2011.

Risk Policy Report. 2011b. Activists Laud HHS' Formaldehyde Cancer Listing Over Industry Objection. Risk Policy Report 18(26). June 28, 2011.

Zhang, L. C. Steinmaus, D.A. Eastmond, X.K. Xin, and M.T. Smith. 2009. Formaldehyde exposure and leukemia: A new meta-analysis and potential mechanisms. Mutat. Res. 681(2-3):150-168.

2

Review of the Formaldehyde Profile in the National Toxicology Program 12th Report on Carcinogens

To address the first part of its statement of task, this committee reviewed the formaldehyde substance profile in the National Toxicology Program (NTP)'s 12th Report on Carcinogens (RoC) (NTP 2011). The committee's review was informed by many documents, including those in Table 1-1. The committee also examined the primary literature cited in the background document for formaldehyde and other literature published by June 10, 2011 (the date when the 12th RoC was released). The headings and structure of the present chapter parallel the major headings that NTP used in the substance profile for formaldehyde— that is, cancer studies in humans, cancer studies in experimental animals, other relevant data, and studies of mechanisms of carcinogenesis. The committee also reviewed the following sections in the substance profile: properties, use, production, exposure, regulations, and guidelines.

As part of its review, the committee determined whether NTP had described and conducted its literature search appropriately, whether the relevant literature identified during the literature search was cited and sufficiently described in the background document, whether NTP had selected the most informative studies in making its listing determination, and whether NTP's arguments supported its conclusion that formaldehyde is known to be a human carcinogen. Instead of discussing the strengths and weaknesses of each study in detail as part of this chapter, the committee chose to discuss such detail as part of its independent analysis in Chapter 3. Detailed data from individual studies can be found in Chapter 3 and in the background document for formaldehyde (NTP 2010). On the basis of its review and analysis of the substance profile, the committee ends this chapter with a review of NTP's literature-search methods, suggestions of clarifications that NTP could make to improve future iterations of the background document or substance profile for formaldehyde, and an assessment of whether the evidence presented by NTP in the background document and the substance profile support the listing of formaldehyde as a known human carcinogen in the 12th RoC.

CARCINOGENICITY

NTP began the substance profile with a clear statement of its conclusions—that is, formaldehyde is known to be a human carcinogen. That conclusion was based on evidence from studies in humans and supporting mechanistic data. The introductory paragraph also informs the reader that formaldehyde was first listed in the 2nd RoC as reasonably anticipated to be a human carcinogen, and that the substance was upgraded to its current listing status of known to be a human carcinogen in the 12th RoC. The committee finds this paragraph to be informative.

Cancer Studies in Humans

The committee reviewed the "Cancer Studies in Humans" section in the NTP substance profile and the corresponding sections in the background document. NTP described the search strategy used to identify relevant epidemiologic studies, and the committee judged the choice of substance-specific and topic-specific terms to be reasonable. The committee did not identify any informative epidemiologic studies that were omitted from the background document. The committee judged that the most informative studies were cited by NTP and were appropriately summarized in the substance profile.

The distinctions among subtypes of nasopharyngeal and sinonasal cancers were adequately discussed in the background document and in the substance profile. The relevance of the subtypes for the determination of carcinogenicity is appropriately discussed. The evidence in the available literature on which subtypes of nasopharyngeal and sinonasal cancers are increased in incidence by exposure to formaldehyde is modest and not definitive. An increase in incidence that is modest and not definitive is not surprising given the rarity of these tumors and the difficulty of having sufficient statistical power to distinguish patterns of association by subtype of cancer (NTP 2010). The limitation in the literature related to cancer subtypes is appropriately discussed in the substance profile and does not materially limit the validity of the carcinogenicity determination.

The committee agrees with NTP's focus on three principal types of cohort and case–control studies in humans: studies of industrial workers, studies of professional groups that have high exposure (embalmers), and studies of general-population cohorts and case–control studies. The first two of those provided the most informative evidence because of greater opportunities for exposure of workers and because of higher-quality exposure assessment as a component of the study method. The committee agrees with NTP's judgment that the two most informative occupational studies for evaluating human cancer hazard posed by formaldehyde are the National Cancer Institute (NCI) study of a cohort of more than 25,000 workers in industries that use formaldehyde (Beane Freeman et al. 2009) and the NCI nested case–control study of cancer in embalmers (Hauptmann et al. 2009). That judgment was based on the strengths of the studies—

they are large, high-quality studies that used well-documented methods and high-discrimination exposure assessments. The committee judged the exposure assessments to be of good quality because they included detailed evaluations of the sources and variations in exposure and used appropriate statistical modeling to estimate unmeasured historical exposures (see Appendix C for more discussion). Additional strengths of the NCI embalmer study are the likelihood of high exposures for long periods, a well-conducted exposure reconstruction, and a careful analysis of alternative measures of quantitative exposure (Hauptmann et al. 2009).

The National Institute for Occupational Safety and Health has produced a mortality study of a cohort of garment workers (Pinkerton et al. 2004). The cohort was relatively small (2,206 total deaths observed over more than 40 years), and the exposure assessment was less detailed, but the likelihood of substantial exposure before 1970 was clearly documented (Elliot et al. 1987), and the study methods and conduct were rigorous. The study was not informative on the question of an association between formaldehyde and nasopharyngeal or sinonasal cancers because of low statistical power. If the cohort had experienced the mortality rates of the general population in the United States, not even one nasopharyngeal cancer death would have been expected in a population of this size. And, consistent with this expectation, no nasopharyngeal cancer deaths were observed. The same is true in this study for sinonasal cancer—less than one death was expected and zero were observed. A British chemical-worker study conducted by Coggon et al. (2003) had a semi-quantitative exposure assessment and was probably also insufficiently powered to determine whether nasopharyngeal carcinoma or sinonasal carcinoma is associated with formaldehyde exposure.

NTP considered several population-based case–control studies to be particularly valuable in the assessment of carcinogenicity. The assessment of formaldehyde exposure and nasopharyngeal cancer was informed by a population-based case–control study by Vaughan et al. (2000) and three smaller case–control studies of nasopharyngeal cancer by Roush et al. (1987), West et al. (1993), and Hildesheim et al. (2001).The assessment of formaldehyde exposure and sinonasal cancer was informed by the pooled case–control studies of sinonasal cancer reported by Luce et al. (2002) and several smaller case–control studies of sinonasal cancer by Olsen et al. (1984), Hayes et al. (1986), Olsen and Asnaes (1986), Roush et al. (1987), and Luce et al. (1993). Because sinonasal cancers are rare (NTP 2010), it was appropriate for NTP to give substantial weight to the findings from the pooled analysis by Luce et al. (2002) of 12 case–control studies, each of which individually lacked sufficient statistical power to detect an effect. The data from those studies could be combined (pooled) because of common methods of data collection and because a detailed exposure reconstruction was conducted specifically for the pooled analysis.

NTP drew on the findings of a meta-analysis by Zhang et al. (2009) and cited meta-analyses by Bachand et al. (2010) and Bosetti et al. (2008). Meta-analyses can be useful in summarizing results of multiple studies, but after reviewing the published meta-analyses on formaldehyde and evaluating their methodologic differences, the committee decided not to use the published meta-analyses or to conduct its own meta-analysis for its independent assessment of formaldehyde in Chapter 3. Because of the considerable heterogeneity in design, particularly among the exposure assessments, the results of a meta-analysis of the full range of observational studies published on formaldehyde exposure and cancer would be highly sensitive to inclusion and exclusion criteria and to other methodologic decisions (Checkoway et al. 2004).

The substance profile described only briefly why Zhang et al. (2009) was given some weight in the assessment of carcinogenicity but Bachand et al. (2010) and Bosetti et al. (2008) were not. Zhang et al. (2009) hypothesized that acute myeloid leukemia (AML) was associated with formaldehyde exposure. The study was unusual in that, unlike some meta-analyses, it had a careful exposure rationale for its approach. The authors decided to focus their analyses by using only the highest exposure categories to obtain the strongest test for a relationship between exposure and disease frequency. They assumed that if an increased frequency of AML was observed, it would most likely be found by analyzing the contrast between the most highly exposed subjects and the unexposed subjects. They argued that higher relative risks were less susceptible to type 2 errors, higher-exposure categories would be less affected by risk dilution by subjects who had low exposures, and high relative risks were less likely to be a result of confounding factors. They also focused on myeloid leukemia instead of all leukemias because they had hypothesized that AML was causally linked to formaldehyde exposure. In the committee's own assessment (described in Chapter 3), no meta-analyses were considered, because they were not deemed necessary in reaching a strong conclusion and because of the difficulties in evaluating conflicting results from different meta-analyses.

The committee concluded that NTP did a thorough job of describing the epidemiology literature in the background document and synthesizing information about key studies in the substance profile. However, the substance profile was not transparent about how the epidemiology evidence met the RoC listing criteria. The listing criteria indicate that formaldehyde should be categorized as reasonably anticipated to be a human carcinogen if "there is limited evidence of carcinogenicity from studies in humans, which indicates that causal interpretation is credible, but that alternative explanations, such as chance, bias, or confounding factors, could not be adequately excluded" (NTP 2010, p. iv). Formaldehyde should be categorized as known to be a human carcinogen if "there is sufficient evidence of carcinogenicity from studies in humans, which indicates a causal relationship between exposure to [formaldehyde]…and human cancer" (NTP 2010, p. iv). There was no discussion in the "Cancer Studies in Humans" section of the background document or substance profile about how NTP defined the terms *limited evidence* and *sufficient evidence*. Therefore, consistent with the RoC listing criteria, the committee used its expert scientific judgment to interpret and apply the listing criteria to the evidence evaluated in Chapters 2 and 3. *Limited evidence* was defined by the committee as evidence from two or more strong or moderately strong studies with varied study designs and populations that suggested an association between exposure to formaldehyde and a specific cancer type, but alternative explanations, such as chance, bias, or confounding factors, could not be adequately ruled out because of limitations in the studies, and so a causal interpretation could not be accepted with confidence. *Sufficient evidence* was defined by the committee as consistent evidence from two or more strong or moderately strong studies with varied study designs and populations that found an association between exposure to formaldehyde and a specific cancer type and for which chance, bias, and confounding factors could be ruled out with reasonable confidence because of the study methodologies and the strength of the findings. The way in which the committee categorized studies as strong, moderately strong, or weak is described in more detail in Chapter 3.

Nasopharyngeal Cancer

As was accurately summarized in the substance profile, nasopharyngeal cancers are a group of uncommon tumors with several histologic types, including differentiated keratinizing squamous-cell carcinoma, differentiated nonkeratinizing carcinoma, and undifferentiated nonkeratinizing carcinoma. NTP based its evaluation of epidemiologic evidence of nasopharyngeal cancer on several lines of evidence. The committee reviewed the background document and the findings of a previous expert panel (McMartin et al. 2009) and concurs with the choice of the key studies presented in the substance profile.

NTP found several case–control studies to be highly informative, notably a case–control study by Vaughan et al. (2000) that drew incident cases from five US cancer registries that participated in the Surveillance Epidemiology and End Results program of NCI (Vaughan et al. 2000). The committee noted two important contributions of the Vaughan et al. (2000) multicenter case–control study: it was able to evaluate risks separately for the three principal types of nasopharyngeal tumors described above, and the exposure assessment was sufficiently detailed to provide evidence of a strong dose–response relationship. Corroborating evidence was provided by additional case–control studies by Roush et al. (1987), West et al. (1993), and Hildesheim et al. (2001).

The NCI industrial worker cohort also provided important corroborating evidence of an association between formaldehyde exposure and nasopharyngeal-cancer mortality, although the rarity of the tumors limited the statistical power of the study (Hauptmann et al. 2004). The authors observed a pattern of increased mortality among categories of exposure defined by duration of exposure, average intensity cumulative exposure, and peak exposure. Although the number of cases was not as large, the study was strengthened by its high-quality exposure assessment. The design of the NCI industrial worker cohort consisted of employees in 10 plants. The objective was to obtain a sufficiently large study group to determine causes of increased mortality for common cancers. However, only a small number of nasopharyngeal-cancer deaths occurred. Of the nine deaths, five occurred in workers in a single plant, which was the second largest plant in the study (Hauptmann et al. 2004). As noted by Hauptmann et al. (2005), it is not unusual to see large variation in small numbers of rare cancers across small plants. Two possible explanations for the heterogeneity in outcomes by plant is the heterogeneity in exposures across the plants and the possibility of confounding by other carcinogenic exposures in the plant that had the most cases. To evaluate that possibility, the investigators conducted analyses that adjusted for the plant. The results of the adjusted analyses were substantially similar to the unadjusted findings, although limited by the small numbers of cases.

Sinonasal Cancer

The committee agreed with NTP's assessment that the Luce et al. (2002) study of sinonasal cancer was particularly useful. As noted, it was a pooled analysis of several high-quality case–control studies that shared the same exposure assessment, and the resulting statistical power was critical for the study's findings. The study found a substantial increase in the frequency of one type of sinonasal cancer—adenocarcinoma—after high cumulative exposure to formaldehyde in both men and women. NTP determined that earlier case–control studies by Olsen et al. (1984), Hayes et al. (1986), Olsen and Asnaes (1986), Roush et al. (1987), and Luce et al. (1993) taken as a group provided consistent supporting evidence of an association.

The committee found that the issue of potential confounding of the formaldehyde–sinonasal-cancer association by wood dust was adequately considered by NTP. The substance profile noted that Hansen and Olsen (1995, 1996) were conducted in occupational cohorts in which wood-dust exposure was very unlikely. In addition, several studies either stratified by likely wood-dust exposure (Olsen et al. 1984; Hayes et al. 1986; Olsen and Asneas 1986) or fitted models to control for confounding by wood dust statistically (Luce et al. 2002). Although each of the studies taken alone had some limitations because of small numbers of cases, on balance the evidence supports NTP's conclusion that the observed association between formaldehyde and sinonasal cancer is unlikely to be due to confounding by wood-dust exposure. No other important confounders were identified in the available studies.

Lymphohematopoietic Cancer

The committee reviewed the background document (NTP 2010) and the findings of the previous expert panel (McMartin et al. 2009) and concurs with the choice of key studies presented under the heading "Lymphohematopoietic Cancer" in the substance profile (NTP 2011). The committee agrees with NTP that the most informative primary studies for evaluating formaldehyde exposures and lymphohematopoietic cancers were the NCI study of the cohort of industrial workers exposed to formaldehyde (Beane Freeman et al. 2009) and the NCI nested case–control study of embalmers (Hauptmann et al. 2009). As previously mentioned, those studies are informative because of their size and the quality of their design and conduct, particularly because the quality of the extensive exposure assessments permitted quantitative evaluations with a variety of plausible exposure metrics. NTP determined that the most informative studies for evaluating formaldehyde exposure and myeloid leukemia specifically were the British cohort of industrial workers (Coggon et al. 2003), the NIOSH cohort of garment workers (Pinkerton et al. 2004), the NCI cohort of industrial workers (Beane Freeman et al. 2009), and the NCI nested case–control study of embalmers (Hauptmann et al. 2009). The epidemiology literature is discussed in more detail in Chapter 3.

The committee found that the assessment of lymphohematopoietic cancers presented in the substance profile supports NTP's conclusion that the most strongly supported association is that between myeloid leukemia and formaldehyde. Broader diagnostic categories (all leukemias and all lymphohematopoietic cancers) also show evidence of an association with formaldehyde exposure in some studies, but a likely explanation for those increases is the inclusion of myeloid leukemia in the broader groupings that include it. The committee agrees with NTP that the evidence demonstrates an association between exposure to high concentrations of formaldehyde (by several different metrics) and some lymphohematopoietic cancers, specifically myeloid leukemia. That association cannot be explained by chance, bias, or confounding factors (NTP 2011).

Lymphohematopoietic cancers make up a diverse group that are often analyzed together in epidemiologic studies because of the rarity of the individual types. Concerns have been raised about the usefulness of such a broad category of tumors when evidence of carcinogenicity is being evaluated because the different cancers of the hematopoietic system are understood to arise from different cells and so might have different etiologic mechanisms (NRC 2011). There is a common assumption in epidemiology, dating back at least to Bradford Hill (Hill 1965), that specific hazardous exposures, such as exposures to chemicals, tend to cause diseases by a small number of specific pathways (or modes of action), so there is an expectation of observing stronger associations between exposures and narrowly defined disease entities than between exposures and broad categories. Nevertheless, it is common practice in epidemiology to begin an evaluation of exposure–disease associations by looking for signals of an association in broad and heterogeneous groups of diseases (including, for example, the very broad category of all cancers combined or all lung diseases combined). If evidence of an association is found in a broad disease category in an exposed population, the next step is to look into more narrowly defined disease subgroups, such as different types of leukemias. Sometimes, the result is that an association is observed only in the broad group and not in any of the constituent disease subgroups. In such cases, the result is spurious and could possibly be explained by bias in study design or data collection. The more likely explanation, though, is that the association observed in the broad disease category might be accounted for by an increase in the association of one or a small number of specific disease entities when the rest of the broad group shows no increase in the association. The latter pattern is interpreted by the study authors as evidence of an association between the exposure and the specific subgroup or subgroups of the disease.

Cancer at Other Tissue Sites

The substance profile discusses cancer at other sites only briefly, so the committee's assessment of this section is based on the review in the background document (which is also brief but more informative) and a review of some of the primary literature. The committee concurred with NTP's assessment that the literature published by June 10, 2011, does not meet the requirement of limited evidence of a carcinogenic effect at any additional sites. As stated in the background document, "in general, the reported estimates were null [relative risk = 1.0] or slightly elevated but statistically nonsignificant, and studies have not consistently reported an elevated risk in cancer associated with formaldehyde exposure at any of these sites" (NTP 2010, p. 232).

Conclusions Regarding Epidemiologic Evidence

The committee concurs with NTP that there is sufficient evidence in studies that had adequate characterization of relevant exposure metrics to enable a conclusion about human cancer after exposure to formaldehyde. The strongest studies are ones that had high-quality exposure assessments and ones that presented alternative exposure metrics. As noted above, there are several such studies. NTP's discussions of chance, bias, confounding factors, and other limitations of the most informative studies in the substance profile are clear and thorough. The committee agrees with NTP's determination that the human evidence published by June 10, 2011, on the association of exposure to formaldehyde with cancer of the nasopharyngeal region and sinonasal cavities and of myeloid leukemia was sufficient to support a listing as known to be a human carcinogen.

Cancer Studies in Experimental Animals

The section "Cancer Studies in Experimental Animals" in the substance profile discusses the degree of certainty of the carcinogenicity of formaldehyde on the basis of evidence from experimental animal studies. According to the NTP listing criteria (Box 1-2), evidence from animal studies is to be judged sufficient to categorize a chemical as reasonably anticipated to be a human carcinogen if "there is increased incidence of malignant and/or a combination of malignant and benign tumors in multiple species or multiple tissue sites; by multiple routes of exposure; or to an unusual degree with regard to incidence, site, or type of tumor, or age at onset" (NTP 2010, p. iv). The committee reviewed the substance profile in the context of those criteria.

Neither the formaldehyde substance profile (NTP 2011) nor the background document (NTP 2010) present details of the approach taken to search the literature for animal carcinogenicity studies although a good description of the literature-search strategy was provided by NTP in response to committee inquiry (Bucher 2013; see Table 2-1). The committee reviewed the comprehensive compilation of animal bioassays in the US Public Health Service 149 series *Survey of Compounds Which Have Been Tested for Carcinogenicity* and evaluations by the International Agency for Research on Cancer (IARC 1982, 1995, 2006), but it did not find other important or informative animal carcinogenesis studies that were missed by NTP and should have been included in the background document or in the substance profile. It found a few early studies of low power (small numbers of animals were used), of poor quality, or of short duration that were not described in the background document. Examples include a 6-month lung exposure study in rabbits that found atypical proliferation (Garschin and Schabad 1936), a study with no controls that administered formaldehyde to 10 rats via subcutaneous injection and found injection-site sarcomas in four (Watanabe et al. 1954), and a 10-month oral experiment in six rabbits that found intraepithelial carcinoma in the exposed mucosa in two (Muller et al. 1978). Those studies contribute little evidence on formaldehyde carcinogenicity, and the RoC and background document are not remiss or deficient for not evaluating them.

Inhalation

In the "Inhalation" section of the formaldehyde background document, studies are grouped by species. Two studies discussed in the background document used mice. The study by Horton et al. (1963) focused on the lung and did not examine the nasal epithelium. C3H mice were exposed to formaldehyde at 0.05 mg/L (50 mg/m^3) for 35 weeks and then for 29 weeks of repeated formaldehyde exposure to 0.15 mg/L (150 mg/m^3), for a total of 64 weeks, and then all mice were sacrificed. None of the mice were found to develop pulmonary neoplasms. The committee judged that this omission from the background document was appropriate given the severe limitations of the study. Furthermore, the study was not noted in the animal-evidence section of the substance profile and, given the limitations of the study, NTP was reasonable to exclude it from further consideration.

Kerns et al. (1983a) conducted a 2-year study of male and female B6C3F$_1$ mice with relatively large dose groups, interim sacrifices (at 6, 12, 18, and 24 months), adequate statistical evaluation, and thorough histopathologic examination of the nasal turbinates and other components of the respiratory tract. Nasal lesions of increasing severity with increasing dose were reported, and two of 17 surviving males in the highest-dose group had squamous-cell carcinoma (Kerns et al. 1983b). The two squamous-cell carcinomas were attributed to formaldehyde given the rareness of the tumors (the background document reported no tumors of this type in 2,800 historical control animals from NTP studies) and the similarity of the lesions observed in rats by the same authors. The substance profile cited that as evidence of carcinogenicity in male mice, and the committee finds this reasonable.

The discussion of the studies in rats in the background document groups the studies as "subchronic" and "chronic". The subchronic-exposure studies (Rusch et al. 1983; Woutersen et al. 1987; Wilmer et al. 1989) might have been more appropriately placed in a section on "other relevant data" that discussed proliferative lesions. The proliferative lesions observed in the short-term studies (for example, squamous-cell metaplasia and hyperplasia of the nasal epithelium) were observed to precede squamous-cell carcinoma in the chronic studies. However, the short-term studies were of insufficient duration to produce tumors and are not themselves carcinogenesis studies. Their exclusion from the substance profile discussion of animal carcinogenesis is appropriate.

The subchronic study by Feron et al. (1988) exposed male Wistar rats to formaldehyde for 13 weeks and then sacrificed the animals after an additional 118 weeks. The 118-week followup period allowed sufficient time for the effects of the 13-week exposure to be manifested. The background document noted the variety of nonneoplastic changes in the olfactory epithelium in addition to the nasal tumors observed (polypoid adenoma, squamous-cell carcinoma, and carcinoma in situ). The fact that tumors developed after short-term exposure was appropriately noted in the substance profile.

The discussion of the chronic rat studies in the background document begins with the large multidose studies in male and female Fischer 344 (F344) rats sponsored by the Chemical Industry Institute of Toxicology (Swenberg et al. 1980a,b; Kerns et al. 1983a). The studies were considered to be state-of-the-art for the time; the methods included a large group, multiple interim sacrifice times, and full histopathologic evaluation of nasal tissue for characterization of neoplastic and nonneoplastic lesions. The studies were well described in the background document. The finding of a high incidence of rare nasal tumors in male and female rats provides a logical and definitive basis for NTP's conclusion on formaldehyde-induced nasal carcinogenesis.

The background document cited additional long-term inhalation-carcinogenesis studies in rats. Woutersen et al. (1989) evaluated the effects of damage to the nasal epithelium in male Wistar rats. During the first week of the study, the nasal mucosa of some rats was severely damaged by electrocoagulation. A higher nasal-tumor incidence was observed in exposed rats that had damaged nasal epithelium than in rats that had undamaged nasal epithelium, although the study of rats with undamaged epithelium had smaller groups (this was not noted in the background document). Monticello et al. (1996) reported on the relationship between indexes of cell proliferation and induction of nasal tumors in relatively large groups (90–147) of male F344 rats that were exposed to a range of concentrations (0.7–15 ppm for 6 hours/day, 5 days/week) for up to 2 years. Squamous-cell carcinoma and polypoid adenoma of the nasal cavity were again found. Kamata et al. (1997) exposed smaller groups of male F344 rats (32 per group), performed histopathologic evaluations of respiratory tract and non–respiratory tract tissue, and similarly found squamous-cell carcinoma of the nasal cavity but no cancers at other sites. Sellakumar et al. (1985) studied the effects of formaldehyde in male Sprague Dawley rats and performed a histopathologic evaluation of other major tissues; the study did not appear to include bone marrow. Nasal squamous-cell carcinoma was found at a relatively high incidence (38% in the group dosed with formaldehyde at approximately 15 ppm). The studies were each adequately described in the background document and reported in the substance profile as providing evidence of formaldehyde-induced carcinogenicity in the nasal epithelium. The committee agrees with the inclusion of the studies because they support the overall sufficiency of evidence of carcinogenicity in animals exposed to formaldehyde.

One chronic study in female Sprague Dawley rats was not cited in the substance profile and was discounted in the background document because of small groups (Holmström et al. 1989). Squamous-cell metaplasia was observed, but only one animal developed nasal squamous-cell carcinoma. The study authors concluded that the finding in the one animal was related to formaldehyde exposure, but NTP did not include that as supportive in the substance profile—a reasonable decision given the observation of a single tumor.

Two monkey studies presented in the background document's table of nasal-tumor results were too short to be reported with other carcinogenesis studies, especially in such long-lived animals. One study was in cynomolgus monkeys and was 26 weeks long (Rusch et al. 1983), and the other study was in rhesus monkeys and was only 6 weeks long (Monticello et al. 1989). The limitation regarding study length was not noted in the study description in the background document but was noted in the summary table of carcinogenicity results. It would have been more appropriate not to include those studies in the section on cancer-bioassay data. They were not mentioned in the substance profile, and that is appropriate.

Two inhalation studies in Syrian golden hamsters are discussed in the background document. One was only 26 weeks in duration, included a small group size (10 male and 10 female), and resulted in no significant findings (Rusch et al. 1983). The background document reported that the study was of short exposure duration and used a small number of animals. The study was not noted in the substance profile and, because it was not a carcinogenesis study, that is appropriate. The study was insufficient as a carcinogenesis study, and NTP would not have been faulted if it had left it out of the background document. The second inhalation study exposed two groups of Syrian golden hamsters over a lifetime (Dalbey 1982). One group (n=88) was exposed 5 hours/day, 5 days/week at 10 ppm, and the second group (n=50) was exposed 5 hours/day, 1 day/week at 30 ppm. Higher incidences of nasal metaplasia and hyperplasia were observed in the 10-ppm group than in the 132 control animals, but no nasal tumors were present. The substance profile did not include the study as a basis of its finding of sufficient evidence for carcinogenesis, and that is appropriate.

Oral

The background document describes studies in which relatively high concentrations of formaldehyde were administered to rats in drinking water. In the first study described, eight of 10 Wistar rats that received 5,000 ppm of formalin in drinking water developed squamous-cell papilloma of the forestomach compared to none of 10 control animals (Takahashi et al. 1986). The finding is noted in the substance profile, and the committee finds that appropriate. Two other drinking-water studies in Wistar rats (Til et al. 1989; Tobe et al. 1989) found epithelial hyperplasia and hyper keratosis of the forestomach and hyperplasia of the glandular stomach, but no statistically significant differences in tumors between the treated and control animals. One of the studies (Til et al. 1989), which had a reasonable size (70 animals/group), exposed male and female rats for up to 2 years to average concentrations of 20, 260, and 1,900 ppm in drinking water. Tobe et al. (1989) designed a 2-year study with concentrations of formaldehyde in drinking water at 0, 200, 1,000, and 5,000 ppm and group sizes of 20 animals of each sex. None of the high-dose animals survived to the end of the study.

The background document describes well the series of drinking-water experiments conducted by Soffritti et al. (1989, 2002) in male and female Sprague Dawley rats. Soffritti et al. (1989) exposed animals in utero (dams exposed via drinking water) and postnatally for 2 years. Breeders were exposed for a lifetime. In the female offspring, the incidence of malignant intestinal tumors was significantly increased. In a statistical analysis of the study, IARC (2006) found that the incidence of intestinal leiomyosarcoma was significantly increased in female offspring and in male and female offspring combined. The substance profile noted that benign and malignant gastrointestinal tumors were reported, including rare intestinal leiomyosarcomas in females. Because leiomyosarcoma is rare, even with the low incidence NTP deemed the finding significant; this is similar to the IARC (2006) conclusion. The substance profile includes the finding and, although the finding is not robust, it is not unreasonable for NTP to include it. In the second series of studies by Soffritti et al. (2002), rats were exposed as adults, and males in the high-dose group were observed to have gastrointestinal leiomyosarcomas, and females in the high-dose group were observed to have leiomyomas. This second finding of gastrointestinal leiomyosarcoma was again given weight because of the rarity of the tumor. In those studies, an increased incidence of hemolymphoreticular tumors was observed, but the finding was not given much weight, because of large discrepancies between the initial incidence reported in a preliminary report and the final published incidence, because of pooling of lymphomas and leukemias, and because limited information was given on the tumor incidence in historical controls. Soffritti et al. (2002) also reported significant increases in tumors of the mammary gland, but the significance did not persist when liposarcomas were removed from the group. The committee agrees with not giving weight to the hemolymphoreticular and mammary tumors in the substance profile and with attaching some weight to the leiomyosarcomas.

Coexposure with Other Substances

The substance profile notes that formaldehyde promotes tumors of the stomach and lung in rats and cites the background document as a reference. There were nine coexposure carcinogenicity studies of varied study design. The results of some studies were null. NTP did not include any of the studies in its findings on the sufficiency of the evidence in animals. That scientific judgment is consistent with the NTP criteria.

Conclusion Regarding Animal Evidence

NTP concluded that the experimental evidence was sufficient to find that formaldehyde is an animal carcinogen. With regard to NTP's application of its criteria, it noted that formaldehyde caused "tumors in two rodent species, at several different tissue sites, and by two different routes of exposure" (NTP 2011, p. 197). A positive finding on any one of the three conditions listed below in which malignant or combined malignant and benign tumors occur would fulfill the criteria for sufficiency in animals. NTP found that two were met.

1. In multiple species or multiple tissue types:
   - Multiple species: NTP cites studies that showed increases in malignant tumors in rats (Feron et al. 1988; Kerns et al. 1983a; Sellakumar et al. 1985; Soffritti et al. 1989; Woutersen et al. 1989; Monticello et al. 1996; Kamata et al. 1997) and in mice (Kerns et al. 1983a).
   - Multiple tissue types: NTP cites studies that showed malignancies of the nasal epithelium (mostly squamous-cell carcinomas) (Kerns et al. 1983a; Sellakumar et al. 1985; Feron et al. 1988; Woutersen et al. 1989; Monticello et al. 1996; Kamata et al. 1997) and gastrointestinal tract (leiomyosarcoma) (Soffritti et al. 1989 [offspring]; Soffritti et al. 2002 [adults]). The substance profile also noted that benign testicular adenoma was seen in the Soffritti et al. (2002) study.

2. After exposure by multiple routes: NTP cites exposure by inhalation (Kerns et al. 1983a; Sellakumar et al. 1985; Woutersen et al. 1987; Feron et al. 1988; Monticello et al. 1996; Kamata et al. 1997) and oral routes (Soffritti et al. 1989 [offspring]; Soffritti et al. 2002).

3. To an unusual degree with respect to incidence, site, type of tumor, or age at onset: NTP did not state that this criterion was met; however, nasal tumors are rarely increased in animal studies, and these tumors were observed at relatively high incidences in the formaldehyde animal studies (Kerns et al. 1983a; Monticello et al. 1996).

The committee agrees with NTP's conclusion that there is sufficient evidence of carcinogenicity in animals to support a listing in the 12th RoC.

Other Relevant Data

The section "Other Relevant Data" of the substance profile presents a selection of studies that deal with formaldehyde and the following topics: chemical reactivity, toxicity (in vivo and in vitro), systemic and organ-specific effects, genomic effects (mutagenic), covalent adducts (protein and DNA), carcinogenicity of formaldehyde metabolites and related compounds, and absorption, distribution, metabolism, and kinetics. Many of the studies are referred to in detail in other sections of the substance profile. The text in this section succinctly and appropriately describes the current understanding of the regional respiratory-tract absorption of formaldehyde and provides appropriate literature citations. The text also accurately describes the reactivity of formaldehyde with water and biologic molecules and accurately indicates the short plasma half-life of formaldehyde. Biomarkers of formaldehyde's interaction with macromolecules (blood proteins and DNA adducts) are well documented. The current understanding of the cytotoxicity of formaldehyde is adequately covered.

Studies on Mechanisms of Carcinogenesis

The section "Studies on Mechanisms of Carcinogenesis" in the substance profile and the associated sections in the background document describe the scientific evidence and mechanistic knowledge available on the carcinogenicity of formaldehyde. The committee finds that the extent, quality, and interpretation of the mechanistic evidence described in these documents are comprehensive and that the importance of this information for the decision to list formaldehyde as a known human carcinogen is clearly explained.

Neither the background document nor the substance profile explicitly describes the literature-search strategy; however, as previously stated, the collection of search terms and other information were available on request from NTP (Bucher 2013). On the basis of this information and the content of the background document and the substance profile, the committee concludes that NTP performed a thorough search and appropriately evaluated studies on mechanisms of carcinogenesis that were published in peer-reviewed sources. The committee concludes that the information presented in the background document and the substance profile is comprehensive, balanced, and inclusive and is accompanied by informative evidence tables and short narratives of individual studies. Summaries were written in a clear manner, and the limitations of the individual studies, where appropriate, are acknowledged and taken into consideration. The mechanistic information provided critical evidence that demonstrates the plausibility of formaldehyde-induced carcinogenesis in both experimental animals and humans. Although there was no clear cutoff date for inclusion of the additional mechanistic studies in the peer-reviewed literature between the time of completion of the background document (November 2009) and the final release of the 12th RoC, the committee concludes that NTP did not miss any publications that had strong mechanistic evidence that would have caused NTP to change the listing of formaldehyde as a known human carcinogen. (See Chapter 3 and Appendix D for more information on the committee's literature search beginning in 2009.)

The substance profile focuses on the mechanisms related to specific clinical sites of cancers, specifically, nasopharyngeal, sinonasal, and lymphohematopoietic cancers. The committee finds that delineation of the available mechanistic evidence into portal-of-entry or systemic effects as defined by NRC (2011) would have made the background document and the substance profile stronger. The mechanisms of carcinogenicity of highly reactive chemicals, including formaldehyde, can differ between portal-of-entry sites and distal sites that their native forms or metabolites might not readily reach. Although there are shortcomings of the evidence described in the section "Cancer at Other Tissue Sites" as acknowledged by NTP, the mechanistic evidence pertaining to the systemic effects of formaldehyde would probably be applicable to any distal tissues.

The committee concludes that NTP correctly states that "the mechanisms by which formaldehyde causes cancer are not completely understood" (NTP 2011, p. 198). There may be several mechanisms of action involved and the mechanisms proposed by NTP are not mutually exclusive and might be related. Although it is clear that the overall strength of evidence differs between the portal-of-entry and systemic health effects, most of the evidence presented in the introductory paragraph in this section focuses on a genotoxic mode of action (NTP 2011). An expert panel that reviewed a draft version of the background document stated that two mechanisms are supported by available evidence in sinonasal–pharyngeal regions where inhaled formaldehyde first comes into contact with the mucous layer of the respiratory tract in mammals (McMartin et al. 2009): a cytotoxicity-induced cellular-proliferation mechanism and a genotoxic mechanism. The information presented in this section appropriately details studies in model organisms and cell-culture systems that provide general evidence applicable to a wide array of human tissues.

Nasal Cancer

Most of the upper aerodigestive tract[1] is directly exposed to formaldehyde when it is inhaled. Various anatomic structures in this region have been identified as potential sites of formaldehyde-associated carcinogenesis in both experimental animals and humans (NTP 2010). This section in the substance profile and the corresponding parts of the background document are comprehensive and

---

[1]The aerodigestive tract is "the combined organs and tissues of the respiratory tract and the upper part of the digestive tracts (including the lips, mouth, tongue, nose, throat, vocal cords, and part of the esophagus and windpipe)" (NCI 2014).

balanced. The committee finds that the information presented in the substance profile agrees with that presented in the background document. Nomenclature of the exact anatomic structures affected by exposure to formaldehyde is important and the section would be clearer if the title reflected the two distinct anatomic sites that have been identified as potential portal-of-entry target sites of formaldehyde carcinogenesis in humans: the nasopharyngeal and sinonasal regions.

Although the emphasis on the various forms of genetic damage observed in the nasal tissue is warranted and the description is comprehensive, the substance profile could have provided a stronger summary of the genotoxic mode of action of formaldehyde in the anatomic sites that come into direct contact with formaldehyde. For example, the nasal passages and surrounding anatomic sites in the upper respiratory tract are affected in rodents (which are obligatory nose-breathers) and humans. However, the oral cavity (for example, the buccal epithelium in exposed humans, who might breathe primarily through the mouth because of irritating effects of formaldehyde on the nasal epithelium) and upper digestive tract (in rodent gavage studies) are also target tissues that come into direct contact with formaldehyde. There is mechanistic evidence of adverse health effects of formaldehyde in those anatomic regions (NTP 2010).

The substance profile identifies several types of genetic damage that have been observed in exposed humans and animal models. They include DNA–protein cross-links, DNA cross-links, nucleotide base adducts and mutations, and micronuclei. Although the description of genetic damage in the substance profile mentioned key findings and cited appropriate references, the topic would benefit from a clear structure and a clear presentation of the evidence similar to the structure and presentation of evidence in the background document. That could be achieved with a tiered presentation of the information, from damage at the level of a nucleotide (for example, adducts and mutations) to that at the level of the DNA structure (for example, cross-links) or chromatin (for example, micronuclei). By focusing on the types of damage and pointing to whether evidence supporting or refuting each type is available from in vitro or ex vivo, animal, or human studies, the substance profile could provide an even more concise and structured description of the plausibility of this mechanism.

Cytotoxicity-induced cellular proliferation is identified as a second plausible mechanism of carcinogenicity of formaldehyde at the portal-of-entry sites. The substance profile presented evidence from studies in rodents that histopathologic lesions in the upper respiratory tract lead to cell proliferation. The committee finds the description and analysis of those studies to be robust and well presented. The substance profile also appropriately points out that several concentration–response studies identified strong concordance between cytotoxicity and proliferation (in subchronic studies) and nasal-tumor incidence (in chronic studies) in rodents.

The substance profile acknowledges that cytotoxicity-induced cellular proliferation has been observed "at anatomical sites that are not thought to be the origin of squamous cell carcinoma" (NTP 2011, p. 199). Although it is not entirely clear what anatomic sites are being referred to here, this subsection correctly points out that this mechanism is not exclusively responsible for formaldehyde's carcinogenicity in the upper respiratory tract, inasmuch as a variety of compounds that alone might induce cell proliferation are known not to pose a cancer hazard in the upper respiratory tract. Those compounds include glutaraldehyde, chlorine, and ethylacrylate (Miller et al. 1985; Wolf et al. 1995; NTP 1999).

The mechanistic studies of the genotoxicity and cytotoxicity of formaldehyde and the later studies of compensatory cell proliferation and apoptosis in the upper aerodigestive tract in rodents have reported effects at concentrations that are within an order of magnitude of human exposures reported in several occupational studies. Whereas few studies involving human subjects have examined cytotoxicity-induced cellular proliferation after exposure to formaldehyde, studies performed with rodent models provide strong mechanistic support for the listing of formaldehyde as a known human carcinogen.

Leukemia

The section "Leukemia" in the substance profile focuses on the systemic effects of formaldehyde at distal sites and specifically on myeloid leukemia. The committee points out that the issue of nomenclature of the anatomic structures affected by exposure to formaldehyde is important, and the section would be clearer if the title was revised to make it clear that the information in it pertains to systemic effects of formaldehyde.

Overall, the substance profile and background document provide a comprehensive and balanced presentation of the evidence pertinent to the effects of formaldehyde at distal sites. It also properly acknowledges the limitations in the current scientific understanding of the mechanisms associated with the plausibility that formaldehyde causes malignancies of the hematopoietic system. The committee finds that the information presented in the substance profile is in agreement with that presented in the background document.

The section "Leukemia" of the substance profile addresses three main issues: the cellular origin of myeloid leukemia, the lack of evidence of systemic distribution of formaldehyde or its metabolites, and a general description of several plausible mechanisms. A brief discussion of the cellular origins of myeloid leukemia frames the challenge that formaldehyde does not seem to reach the bone marrow, where most known leukemogens have been shown to affect hematopoietic progenitor cells. However, there might be indirect mechanisms by which formaldehyde affects bone marrow and circulating cells (see Chapter 3). The committee finds this logic to be reasonable. The substance profile acknowledges that there is little evidence that formaldehyde or its metabolites would reach systemic circulation or tissues other than those in direct contact with the agent. Several key studies have evaluated blood concentrations of formaldehyde after exposure of humans and laboratory animals but found no measurable increases (Heck et al. 1985; Casanova et al. 1988; Heck and Casanova 2004). And a study in rats that used ^{13}C-labeled formaldehyde and evaluated DNA-adduct formation in the nasal epithelium and distal anatomical sites, including the bone marrow, was also acknowledged in the substance profile (but not in the background document) to support the assertion that there is an apparent lack of systemic distribution of inhaled formaldehyde (Lu et al. 2010). One additional study (Moeller et al. 2011) that examined the presence of formaldehyde-associated endogenous and exogenous N^2-hydroxymethyl-dG adducts in nasal mucosa and bone marrow DNA of cynomolgus macaques exposed to ^{13}C-labeled formaldehyde was published within months of the release of the 12th RoC and was not referred to in the substance profile. The committee finds that the information presented in the study was consistent with the evidence presented by Lu et al. (2010) and the arguments that were already laid out in the substance profile; inclusion of the new publication in the 12th RoC would not have changed the overall conclusions.

Given the uncertainties in the scientific understanding of the potential mechanisms of the systemic effects of formaldehyde, the committee finds that NTP could have explicitly acknowledged, as stated in a previous expert panel's report (McMartin et al. 2009), that "while it would be desirable to have an accepted mechanism that fully explains the association between formaldehyde exposure and distal cancers, the lack of such mechanism should not detract from the strength of the epidemiological evidence that formaldehyde causes myeloid leukemia" (p. 28).

Systemic Effects Observed after Inhalation or Oral Exposure

The section "Systemic Effects Observed after Inhalation or Oral Exposure" in the substance profile describes several additional lines of evidence that support the notion that formaldehyde has systemic adverse health effects. Such evidence includes data demonstrating toxicity, genotoxicity, and increased incidence of malignancies at distal sites (NTP 2011) following inhalation of formaldehyde. This section in the substance profile and the corresponding parts of the background document are comprehensive and balanced. The committee finds that the information presented in the substance profile agrees with that presented in the background document. Studies presented in this section are highly informative and argue that although it is yet to be established how formaldehyde can exert adverse effects systemically, the strongest evidence of a systemic effect of formaldehyde is evidence of genotoxicity in blood cells that circulate beyond the portal of entry.

The committee concludes that the study by Lu et al. (2010) and other supporting studies strongly argue for the lack of systemic distribution of inhaled formaldehyde. However, it also concludes that the relevance of formaldehyde-induced DNA adducts to formaldehyde-induced carcinogenesis is uncertain given that the background concentrations of these adducts formed by endogenous exposure to formaldehyde are greater than those induced by exogenous formaldehyde at carcinogenic doses, and that tissue concentrations of the adducts vary within and among species tested. In that regard, the committee highlights a point from a previous expert panel's report that chromosome aberrations are an important biomarker of human cancer (McMartin et al. 2009). The chromosomal aberrations observed in lymphocytes of exposed human subjects constitute strong evidence of potentially genotoxic effects of formaldehyde in circulating blood cells. As acknowledged in the substance profile, evidence of genotoxicity of formaldehyde is extensive. Studies that successfully detected DNA–protein cross-links, strand breaks, micronuclei, and chromosomal aberrations in the circulating blood cells of exposed human subjects are convincing and reproducible. No one study performed with human subjects can establish that formaldehyde is the sole genotoxic agent that caused the observed effects, but the diversity of studies, populations, and exposure scenarios gives strong credence to the overall conclusion. Studies of such effects in experimental animals are less consistent, and the substance profile rightly states that "most [experimental animal] studies found no cytogenetic effects" (NTP 2011, p. 199).

The background document and substance profile also note that toxicity of formaldehyde has been reported to occur in the liver, testes, central nervous system, and other organs that would suggest a systemic effect. The publications that were evaluated by NTP include case reports of humans who ingested formaldehyde, reports of epidemiologic studies of occupational cohorts, and reports of in vivo exposures of experimental animals (rats and mice) of varied duration and dosage. Although the evidence presented in those studies is diverse and credible, NTP correctly states that "the mechanisms for systemic toxicity…are not known" (NTP 2011, p. 199).

Theoretical Mechanisms for the Distribution of Formaldehyde to Distal Sites

The substance profile accurately describes the *theoretical* possibility that formaldehyde might diffuse through nasal epithelia to the bloodstream and then throughout the body. The section also provides appropriate literature citations for the information that is presented. Clearly expressed is the salient issue that because of the reversible nature of formaldehyde's reaction with water (which forms methanediol) or macromolecules, it is theoretically possible that a formaldehyde or methanediol molecule might move throughout the body. Moreover, as appropriately noted in the background document, mathematical simulation modeling efforts that incorporate formaldehyde–methanediol kinetics suggest that formaldehyde might penetrate to the bloodstream in the nose (Georgieva et al. 2003); this raises the possibility that inhaled formaldehyde might reach the systemic circulation.

The section "Theoretical Mechanisms for Distribution to Distal Sites" of the substance profile is narrowly focused. Although it is theoretically possible that formaldehyde might distribute away from the portal of entry to distant tissues, the evaluation in the substance profile would be more complete if the potential for formaldehyde to move throughout the body were discussed in the context of the large amounts of endogenous formaldehyde that are present. Such an evaluation would broaden the discussion from one of the theoretical possibility of systemic distribution to a more precise evaluation of whether it is likely to occur to any important extent. Published data that were not cited in this section of the substance profile indicate that inhaled formaldehyde does not increase blood formaldehyde to concentrations that are substantially above endogenous concentrations (Heck et al. 1985; Casanova et al. 1988; Lu et al. 2010; Moeller et al. 2011). Moreover, large amounts of formaldehyde have not been shown to penetrate to tissues distant from the portal of entry. See the detailed toxicokinetics discussion in Chapter 3.

Other Potential Mechanisms of Formaldehyde-Induced Leukemia

The section "Other Potential Mechanisms of Formaldehyde-Induced Leukemia" in the substance profile offers two additional potential mechanisms to explain formaldehyde-induced leukemia. The first suggested mechanism is that "formaldehyde could damage stem cells circulating in the blood, which travel to the bone and become initiated leukemia cells," and the second is that formaldehyde "could damage stem cells that reside in the nasal turbinates or olfactory mucosa" (NTP 2011, pp. 199-200). Both mechanisms are related to potential direct damage to hematopoietic stem cells in the nasal circulation or nasal mucosa. Literature was cited to support the implicit hypothesis that formaldehyde-induced damage occurs to hematopoietic stem cells at the portal of entry. However, this hypothesis has not been proved experimentally (reported data are related to formaldehyde-induced damage in lymphocytes, not stem cells, in circulation or in nasal mucosa). Thus, this section might appear to provide evidence to support the listing (even using the term *support* twice) although it simply suggests some feasibility of the mechanisms. In the absence of direct evidence, these potential mechanisms do not explain how formaldehyde causes leukemia.

Supporting and critical literature are mentioned appropriately in the background document and substance profile. Because this section does not bear on the listing of formaldehyde and because there is no direct evidence of the mechanisms, the review of the literature and the discussion are appropriately brief.

Hematotoxicity

In the section "Hematotoxicity" of the substance profile, NTP reviews evidence of formaldehyde-induced hematologic effects (NTP 2011). The term *hematotoxicity* might imply a health effect that is not addressed in the studies cited in the substance profile. Indeed, the studies presented in the listing profile demonstrate changes in blood-cell number or function but do not address whether the changes have consequences for the health of the animal or human (for example, autoimmunity, infection, bleeding, or leukemia).

The substance profile cites two studies that support the hypothesis that formaldehyde induces hematologic effects. Substantial space is given to Zhang et al. (2010), who investigated occupational exposure to formaldehyde, hematotoxicity, and leukemia-specific chromosomal changes in cultured myeloid progenitor cells. However, several others studies cited in the background document could also contribute to this topic in the substance profile. For example, Ying et al. (1999) investigated lymphocyte subsets and sister-chromatid exchanges in students exposed to formaldehyde vapor. The authors established some specificity of the hematologic effects of formaldehyde, so citing their study in the substance profile would have strengthened the discussion in this section. Some balance is achieved in the first two sentences of the section in the substance profile although no clear synthesis of the evidence is presented. Because a number of studies provide direct and indirect evidence relevant to this topic, a balanced summary sentence on the overall weight of evidence would be helpful. It is important to note that observed changes in hematopoietic cell number or function do not directly support a mechanism of leukemogenesis but rather establish that formaldehyde has effects either directly or indirectly on hematopoietic cells in the circulation. For clarity, it should be stated how this section affects NTP's listing of formaldehyde as a carcinogen.

PROPERTIES

The section "Properties" of the substance profile details major physicochemical characteristics of formaldehyde. Chemical stability, reactivity, and flammability characteristics are also provided. Overall, this brief section serves its purpose well and provides all necessary information on the chemical itself. The section also includes information on various alternative states of formaldehyde, including a monomeric hydrate methylene glycol (methanediol) form of formaldehyde in dilute aqueous solutions, a solid form (1,3,5-trioxane), and various polymers of eight to 100 formaldehyde units that form paraformaldehyde.

USE

The section "Use" of the substance profile and related background document provide a comprehensive review of industrial uses of formaldehyde and paraformaldehyde. Formaldehyde is used primarily in the production of polymer products and resins, so humans might come into contact with formaldehyde through a variety of consumer products and manufacturing processes. Overall, this section supports well the reasoning for considering inclusion of formaldehyde in the RoC in that it is clear that "a significant number of persons residing in the United State are exposed" (NTP 2010, p. 3) to this chemical.

PRODUCTION

The section "Production" of the substance profile covers the chemical processes used to manufacture formaldehyde and provides quantitative estimates of domestic production and of import and export volumes. Formaldehyde is a high-volume production chemical, and manufacturing of this compound is increasing. This section and corresponding information from the background document supports a potential wide exposure to formaldehyde in the United States, inasmuch as about 30 lb of formaldehyde was produced per person in the United States in the middle 2000s. Much of that formaldehyde is used to manufacture a wide variety of products and it enters the market as a component of industrial resins, building materials, home and office furnishings, mortuary chemicals and preservatives, disinfectants in farming, and consumer products. Substantial quantities are also produced from natural sources and combustion sources. This section supports the inclusion of formaldehyde in the RoC.

EXPOSURE

The goal of the section "Exposure" in the substance profile is to show that there is widespread occupational and general population exposure. The section is divided into two subsections: environmental exposures and occupational exposures. The section on "Human Exposure" in the background document has a subsection on "Biological Indices of Exposure". That subsection is brief and does not consider effects of endogenous formaldehyde formation, which will limit the utility of a biomarker because the variation in endogenous formaldehyde will obscure the small signal produced by exogenous exposure. However, it seems to show that some biomarkers distinguish between exposed and nonexposed workers when the exposure is high enough.

The committee observed that the purpose of the section "Exposure" in the background document was not to critically evaluate the industrial exposures that were present for epidemiologic studies evaluated in the section "Cancer Studies in Humans". Instead, it catalogs the highly heterogeneous data gathered in studies of a wide array of environmental and occupational exposure settings and establishes that substantial occupational exposures and widespread exposures of the general population occur.

REGULATIONS AND GUIDELINES

The "Regulations" and "Guidelines" sections of the substance profile provides a comprehensive list of various rules, regulations, and advisory notices that pertain to formaldehyde. It is clear that many government agencies in the United States have set quantitative limits of exposure in various scenarios and regulate production, use, distribution, and disposal of formaldehyde, but the level of detail provided on these in the background document and substance profile varies widely, and it is not clear in many cases whether the appropriate source can be easily found. Many regulations are dated without links to the appropriate document sources.

REVIEW OF NTP'S LITERATURE-SEARCH METHODS

NTP conducted several literature searches to identify carcinogenicity studies that inform the assessment of formaldehyde in the NTP 12th RoC, and some of that information is presented in the section "Human Cancer Studies" of the background document (NTP 2010). For that specific section in the background document, NTP identified some of its search terms, the databases searched, and the inclusion and exclusion criteria that were used. Such details were not included in the background document for other topics, including studies in experimental animals and mechanistic data.

In response to a request from the committee, NTP provided additional information on its literature search methods (Bucher 2013). PubMed, Scopus, and Web of Science were searched by using substance-specific terms (that is, the substance name, major synonyms, and major metabolites) and topic-specific terms (see Table 2-1). The results underwent a first level of review, during which titles and abstracts were screened for relevance, followed by a second level of review in which the full text of references was reviewed for relevance and substance. In the second level of review, 1,170 references were considered. Some 38 additional references were recommended to NTP by an expert panel (McMartin et al. 2009, 2010). In total, 798 references were cited in the final background document. The date when the searches were run and the specific search strings used for each database were not provided to the committee. The committee found that including more detail on the search strategies and on the inclusion and exclusion criteria would have improved transparency of the methods that NTP used to identify and evaluate relevant scientific literature related to formaldehyde exposure and carcinogenicity. Other committees of the National Academies (IOM 2011; NRC 2011, 2014) have made related recommendations about clearly and concisely describing literature searches, and approaches that ensure greater transparency in literature searches and systematic reviews are being initiated by the US Environmental Protection Agency's Integrated Risk Information System (EPA 2013) and NTP's Office of Health Assessment and Translation (NTP 2013).

The final background document summarizes the literature up to the date of the peer review of the background document (November 2009), and the substance profile includes literature up to the date of the peer review by the NTP Board of Scientific Counselors (June 2010) (Bucher et al. 2013). (see Figure 1-1

TABLE 2-1 Topic-Specific Search Terms Used in NTP's Database Searches

Human Cancer	Animal Tumors	Genotoxicity	ADME and Mechanisms

MeSH terms
  Case reports
  Case–control studies
  Cohort studies
  Epidemiology
  Epidemiologic studies
  Mortality
  Neoplasms
  Occupational exposure
  Prospective studies
  Retrospective studies
  Manpower
 	
Text words
  Case-referent
  Cancer
  Carcinogenic
  Epidemiolog*
  Tumor
  Workers
	MeSH terms
	  Adenocarcinoma
	  Adenoma
	  Carcinogens
	  Carcinoma
	  Neoplasms
	  Precancerous condition
	  Sarcoma
	  Animals
	
	Text words
	  Cancer
	  Foci
	  Malignan*
	  Mice
	  Oncogenic*
	  Rats
	  Tumor
	  Tumorigenic*
		MeSH terms
		  Aneuploidy
		  Cell transformation, neoplastic
		  Chromosome aberrations
		  Cytogenic analysis
		  DNA adducts
		  DNA damage
		  DNA repair
		  Germ-line mutation
		  Micronuclei
		  Mutagens
		  Mutagenesis
		  Mutation
		  Oncogenes
		  Polyploidy
		  Sister chromatid exchange
		  SOS response
		
		Text words
		  Chromosom*
		  Clastogen*
		  Genetic toxicology
		  Strand break
		  Unscheduled DNA synthesis
			MeSH terms
			  Absorption
			  Biotransformation
			  Metabolism
			  Pharmacokinetics
			  Cytochrome P-450 enzyme system
			
			Text words
			  Activation
			  Bioactivation
			  Clearance
			  Detoxif*
			  Distribution
			  Excretion
			  Kinetics
			  Mechanism
			  Metabolite

*The asterisk, sometimes referred to as a "wildcard", represents a truncation and it is used to find all terms that begin with the given text string. Abbreviations: ADME, absorption, distribution, metabolism, and excretion; MeSH, medical subject headings. Source: Bucher 2013.

for a schematic of the 12th RoC process.) NTP periodically reviewed the scientific literature up to the release of the 12th RoC (June 2011) "for any new studies that would warrant a re-review of the NTP's preliminary recommendations to the HHS Secretary for the listing status of formaldehyde" (Bucher 2013). Describing that process in greater detail in the background document, including specific dates, would have added transparency to the development of the background document and substance profile.

SUGGESTED REVISIONS FOR FUTURE EDITIONS OF THE FORMALDEHYDE LISTING IN THE REPORT ON CARGINOGENS

Through its review of the background document and substance profile for formaldehyde, the committee identified several revisions that could be made to improve the formaldehyde listing in future iterations of the RoC (see Table 2-2). Addressing the suggestions in Table 2-2 would add clarity and improve the presentation of information in NTP's assessment of formaldehyde, but making the revisions would not change the overall conclusion of carcinogenicity presented in the substance profile.

SUMMARY AND CONCLUSIONS

In response to the statement of task, the committee examined the substance profile published by NTP as part of the 12th RoC. It also examined supporting documents, including those presented in Table 1-1, and relevant primary literature. The committee considered information presented in review articles, reviews completed by such scientific bodies as IARC, and materials submitted to it by the public.

The committee found that the background document describes the strengths and weaknesses of relevant studies in a way that is consistent and balanced. The substance profile appropriately cites studies showing positive associations that support the listing. However, the substance profile would be more complete if it included more discussion on why weaker, uninformative, inconsistent, or conflicting evidence did not alter NTP's conclusions. Although the committee identified that as a limitation in the substance profile, it would not change NTP's final conclusions as presented in the substance profile.

The committee concludes that NTP comprehensively considered available evidence and applied the listing criteria appropriately in reaching its conclusion. The 12th RoC states that "formaldehyde is known to be a human carcinogen based on sufficient evidence of carcinogenicity from studies in humans and supporting data on mechanisms of carcinogenesis" (NTP 2011, p. 195). The committee agrees with NTP's conclusion, which is based on evidence published by June 10, 2011, that formaldehyde is a known human carcinogen.

TABLE 2-2 Suggested Revisions for the Formaldehyde Substance Profile and Background Document In Future Editions of the Report on Carcinogens

Sections in the Substance Profile for Formaldehyde	Suggested Revisions
Study Identification	• Describe the process for identifying relevant literature (including databases searched, keywords used, and search dates).
Cancer Studies in Humans	• Explicitly define the way in which RoC listing criteria terms such as *limited* and *sufficient* were used in the evaluation of the human studies. • Clarify how the quality and relevance of meta-analyses were evaluated and how and why meta-analyses were included in the assessment of the epidemiology evidence. • Add a more detailed description of how exposure assessments were used to evaluate the evidence from individual epidemiology studies. • Include an explanation of the logic used to decide which tumor groupings or end points to include in evaluating epidemiologic evidence.
Cancer Studies in Experimental Animals	• Consider including a finding that formaldehyde induces tumors to an unusual degree (high incidences of rare squamous-cell tumors of the nasal epithelium).
Other Relevant Data	• Include a description of the portal-of-entry toxicity of formaldehyde. • Add a discussion in the background document for formaldehyde of the extensive metabolism of formaldehyde at the portal of entry. • Add a discussion of formaldehyde as a well-established irritant that has the potential to produce an allergic response.
Studies on Mechanisms of Carcinogenesis	Nasal Cancer • Change the title of the section on "Nasal Cancer" to reflect the two distinct potential portal-of-entry target sites. • Strengthen the summary of the genotoxic mode of action discussion. • Restructure the discussion in the substance profile to parallel the presentation of information in the background document.

Leukemia	• Change the title of the section on "Leukemia" to make it clear that the information pertains to potential systemic effects of formaldehyde. • Explicitly acknowledge that "while it would be desirable to have an accepted mechanism that fully explains the association between formaldehyde exposure and distal cancers, the lack of such mechanism should not detract from the strength of the epidemiological evidence that formaldehyde causes myeloid leukemia" (McMartin et al. 2009). • Discuss the potential for formaldehyde to move throughout the body in the context of the large amounts of endogenous formaldehyde that are present. • Make it clear that the section on "Other Potential Mechanisms of Formaldehyde-induced Leukemia" is intended to show feasibility, not evidence of or support for the mechanisms. • Change the title of the section "Hematotoxicity" to "Hematologic and Immunologic Effects" so that the substance profile is consistent with the background document. In addition, add a balanced summary sentence to that section on the overall strength of the evidence and state how that section affects NTP's listing for formaldehyde as a carcinogen.
Exposure	• Integrate information from the section "Exposure" about environmental and occupational settings into the assessment of the epidemiologic studies in the "Human Studies" section. • Strengthen and focus the listing of heterogeneous data in the background document by removing any incomplete and limited data and by providing a more organized presentation of the information. • Compare and contrast different types of industries in a quantitative manner. Situations that involve exposure to particulate materials that contain formaldehyde could be treated separately. Occupational and other activities that produce peak exposures could also be noted and measured. Time trends, if any, in exposure could be identified. • Adopt a consistent unit of exposure for occupational and environmental exposures.
Regulations and Guidelines	• Provide more information on regulations and include proper references to the sources.

REFERENCES

Bachand, A.M., K.A. Mundt, D.J. Mundt, and R.R. Montgomerty. 2010. Epidemiological studies of formaldehyde exposure and risk of leukemia and nasopharyngeal cancer: A meta-analysis. 40(2):85-100.

Beane Freeman, L.E., A. Blair, J.H. Lubin, P.A. Stewart, R.B. Hayes, R.N. Hoover, and M. Hauptmann. 2009. Mortality from lymphohematopoietic malignancies among workers in formaldehyde industries: The National Cancer Institute Cohort. J. Natl. Cancer Inst. 101(10):751-761.

Bosetti, C., J.K. McLaughlin, R.E. Tarone, E. Pira, and C. La Vecchia. 2008. Formaldehyde and cancer risk: A quantitative review of cohort studies through 2006. Ann. Oncol. 19(1):29-43.

Bucher, J.R. 2013. Follow-up Questions. Material submitted by the NAS Committee on Review of the Formaldehyde Assessment in the NTP 12th RoC and the NAS Committee on Review of the Styrene Assessment in the NTP 12th RoC, April 2, 2013.

Casanova, M., H.D. Heck, J.I. Everitt, W.W. Harrington Jr., and J.A. Popp. 1988. Formaldehyde concentrations in the blood of rhesus monkeys after inhalation exposure. Food Chem. Toxicol. 26(8):715-716.

Checkoway, H., N. Pearce, and D. Kribel. 2004. Research Methods in Occupational Epidemiology, 2nd Ed. Oxford: Oxford University Press.

Coggon, D., E.C. Harris, J. Poole, and K.T. Palmer. 2003. Extended follow-up of a cohort of British chemical workers exposed to formaldehyde. J. Natl. Cancer Inst. 95(21):1608-1615.

Dalbey, W.E. 1982. Formaldehyde and tumors in hamster respiratory tract. Toxicology 24(1):9-14.

Elliot, L.J., L.T. Stayner, L.M. Blade, W. Helperin, and R. Keenlyside. 1987. Formaldehyde Exposure Characterization in Garment Manufacturing Plants: A Composite Summary of Three in-depth Industrial Hygiene Surveys. Division of Surveillance, Hazard Evaluations and Field Studies, National Institute for Occupational Safety and Health, Cincinnati, OH.

EPA (U.S. Environmental Protection Agency). 2013. Part 1. Status of Implementation of Recommendations. Materials Submitted to the National Research Council, by Integrated Risk Information System Program, U.S. Environmental Protection Agency, January 30, 2013 [online]. Available: http://www.epa.gov/iris/pdfs/IRIS%20Program%20Materials%20to%20NRC_Part%201.pdf [accessed Jan. 13, 2014].

Feron, V.J., J.P. Bruyntjes, R.A. Woutersen, H.R. Immel, and L.M. Appelman. 1988. Nasal tymours in rats after short-term exposure to a cytotoxic concentration of formaldehyde. Cancer Lett. 39(1):101-111.

Garschin, W.G., and L.M. Schabad. 1936. About atypical proliferation of the bronchial epithelium with the introduction of formalin into the lung tissue [in German]. Z. Krebsforsch. 43(1):137-145.

Georgieva, A.V., J.S. Kimbell, and P.M. Schlosser. 2003. A distributed-parameter model for formaldehyde update and disposition in the rat nasal lining. Inhal. Toxicol. 15(14):1435-1463.

Hansen, J., and J.H. Olsen. 1995. Formaldehyde and cancer morbidity among male employees in Denmark. Cancer Causes Control 6(4):354-360.

Hansen, J., and J.H. Olsen. 1996. Occupational exposure to formaldehyde and risk of cancer [in Danish]. Ugeskr. Laeger. 158(29):4191-4194.

Hauptmann, M., J.H. Lubin, P.A. Stewart, R.B. Hayes, and A. Blair. 2004. Mortality from solid cancers among workers in formaldehyde industries. Am. J. Epidemiol. 159(12):1117-1130.

Hauptmann, M., J.H. Lubin, P.A. Stewart, R.B. Hayes, and A. Blair. 2005. Re: Mortality from Solid Cancers among Workers in Formaldehyde Industries. The Authors Reply. Am. J. Epidemiol. 161(11):1090-1091.

Hauptmann, M., P.A. Stewart, J.H. Lubin, L.E. Beane Freeman, R.W. Hornung, R.F. Herrick, R.N Hoover, J.F. Fraumeni Jr., A. Blair, and R.B. Hayes. 2009. Mortality from lymphohematopoietic malignancies and brain cancer among embalmers exposed to formaldehyde. J. Natl. Cancer Inst. 101(24):1696-1708.

Hayes, R.B., J.W. Raatgever, A. de Bruyn, and M. Gerin. 1986. Cancer of the nasal cavity and paranasal sinuses, and formaldehyde exposure. Int. J. Cancer 37(4):487-492.

Heck, H.D., and M. Casanova. 2004. The implausibility of leukemia induction by formaldehyde: A critical review of the biological evidence on distant-site toxicity. Regul. Toxicol. Pharmacol. 40(2)92-106.

Heck, H.D., M. Casanova-Schmitz, P.B. Dodd, E.N. Schachter, T.J. Witek, and T. Tosun. 1985. Formaldehyde (CH2O) concentrations in the blood of humans and Fischer-344 rats exposed to CH2O under controlled conditions. Am. Ind. Hyg. Assoc. J. 46(1):1-3.

Hildesheim, A., M. Dosemeci, C.C. Chan, C.J. Chen, Y.J. Cheng, M.M. Hsu, I.H. Chen, B.F. Mittl, B. Sun, P.H. Levine, J.Y. Chen, L.A. Brinton, and C.S. Yang. 2001. Occupational exposure to wood, formaldehyde, and solvents and risk of nasopharyngeal carcinoma. Cancer Epidemiol. Biomarkers Prev. 10(11):1145-1153.

Hill, A.B. 1965. The environment and disease: Association or causation? Proc. R. Soc. Med. 58(5):295-300.

Holmström, M., B. Wilhelmsson, and H. Hellquist. 1989. Histological changes in the nasal mucosa in rats after long-term exposure to formaldehyde and wood dust. Acta Otolaryngol. 108(3-4):274-283.

Horton, A.W., R. Tye, and K.L. Stemmer. 1963. Experimental carcinogenesis of the lung. Inhalation of gaseous formaldehyde or an aerosol coal tar by C3H mice. J. Natl. Cancer Inst. 30:31-43.

IARC (International Agency for Research on Cancer). 1982. Chemicals, Industrial Processes and Industries Associated with Cancer in Humans: An Updating of IARC Monographs Volumes 1 to 29. IARC Monographs on the Evaluation of the Carcinogenic Risks to Humans Supplement 4. Lyon, France: IARC [online]. Available: http://monographs.iarc.fr/ENG/Monographs/suppl4/Suppl4.pdf [accessed June 10, 2013].

IARC (International Agency for Research on Cancer). 1995. Wood Dust and Formaldehyde. IARC Monographs on the Evaluation of the Carcinogenic Risks to Humans Vol. 62. Lyon, France: IARC [online]. Available: http://monographs.iarc.fr/ENG/Monographs/vol62/mono62.pdf [accessed June 10, 2013].

IARC (International Agency for Research on Cancer). 2006. Formaldehyde, 2-Butoxyethanol and 1-tert-Butoxy-propan-2-ol. IARC Monographs on the Evaluation of the Carcinogenic Risks to Humans Vol. 88. Lyon, France: IARC [online]. Available: http://monographs.iarc.fr/ENG/Monographs/vol88/mono88.pdf [accessed June 10, 2013]

IOM (Institute of Medicine). 2011. Finding What Works in Health Care: Standards for Systematic Reviews. Washington, DC: National Academies Press.

Kamata, E., M. Nakadate, O. Uchida, Y. Ogawa, S. Suzuki, T. Kaneko, M. Saito, and Y. Jurokawa. 1997. Results of a 28-month chronic inhalation toxicity study of formaldehyde in male Fisher-344 rats. J. Toxicol. Sci. 22(3):239-254.

Kerns, W.D., K.L. Pavkov, D.J. Donofrio, E.J. Gralla, and J.A. Swenberg. 1983a. Carcinogenicity of formaldehyde in rats and mice after long-term inhalation exposure. Cancer Res. 43(9):4382-4392.

Kerns, W.D., D.D. Donofrio, and K.L. Pavkov. 1983b. The chronic effects of formaldehyde inhalation in rats and mice: A preliminary report. Pp. 111-131 in Formaldehyde Toxicity, J.E. Gibson, ed. Washington, DC: Hemisphere Publishing.

Lu, K., L. Collins, H. Ru, E. Bermudez, and J. Swenberg. 2010. Distribution of DNA adducts caused by inhaled formaldehyde is consistent with induction of nasal carcinoma but not leukemia. Toxicol. Sci. 116(2):441-451.

Luce, D., M. Gerin, A. Leclerc, J.F. Morcet, J. Brugere, and M. Goldberg. 1993. Sinonasal cancer and occupational exposure to formaldehyde and other substances. Int. J. Cancer. 53(2):224-231.

Luce, D., A. Leclerc, D. Begin, P.A. Demers, M. Gerin, E. Orlowski, M. Kogevinas, S. Belli, I. Bugel, U. Bolm-Audorff, L.A., Brinton, P. Comba, L. Hardell, R.B. Hayes, C. Magnani, E. Merler, S. Preston-Martin, T.L. Vaughan, W. Zheng, and P. Boffetta. 2002. Sinonasal cancer and occupational exposures: A pooled analysis of 12 case-control studies. Cancer Causes Control 13(2):147-157.

McMartin, K.E., F. Akbar-Khanzadeh, G.A. Boorman, A. DeRoos, P. Demers, L. Peterson, S.M. Rappaport, D.B. Richardson, W.T. Sanderson, and M.S. Sandy. 2009. Part B – Recommendation for Listing Status for Formaldehyde and Scientific Justification for the Recommendation. Formaldehyde Expert Panel Report [online]. Available: http://ntp.niehs.nih.gov/ntp/roc/twelfth/2009/November/FA_PartB.pdf [accessed July 17, 2013].

McMartin, K.E., F. Akbar-Khanzadeh, G.A. Boorman, A. DeRoos, P. Demers, L. Peterson, S.M. Rappaport, D.B. Richardson, W.T. Sanderson, and M.S. Sandy. 2010. Part A – Peer Review of the Draft Background Document on Formaldehyde. Formaldehyde Expert Panel Report [online]. Available: http://ntp.niehs.nih.gov/ntp/roc/twelfth/2009/november/fa_parta.pdf [accessed July 17, 2013].

Miller, R.R., J.T. Young, R.J. Kociba, D.G. Keyes, K.M. Bodner, L.L. Calhoun, and J.A. Ayres. 1985. Chronic toxicity and oncogenicity bioassay of inhaled ethyl acrylate in Fischer 344 rats and B6C3F1 mice. Drug Chem. Toxicol. 8(1-2):1-42.

Moeller, B.C., K. Lu, M. Doyle-Eisele, J. McDonald, A. Gigliotti, and J.A. Swenberg. 2011. Determination of N2-hydroxymethyl-dG adducts in the nasal epithelium and bone marrow of nonhuman primates following 13CD2-formaldehyde inhalation exposure. Chem. Res. Toxicol. 24(2):162-164.

Monticello, T.M., K.T. Morgan, J.I. Everitt, and J.A. Popp. 1989. Effects of formaldehyde gas on the respiratory tract of rhesus monkeys. Pathology and cell proliferation.. Am. J. Pathol. 134(3):515-527.

Monticello, T.M., J.A. Swenberg, E.A. Gross, J.R. Leininger, J.S. Kimbell, S. Seilkop, T.B. Starr, J.E. Gibson, and K.T. Morgan. 1996. Correlation of regional and nonlinear formaldehyde-induced nasal cancer with proliferating populations of cells. Cancer Res. 56(5):1012-1022.

Muller, P., G. Raabe, and D. Schumann. 1978. Leukoplakia induced by repeated deposition of folmalin in rabbit oral mucosa. Long –term experiments with a new 'oral tank". Exp. Pathol. 16(1-6):36-42.

NCI (National Cancer Institute). 2014. NCI Dictionary of Cancer Terms [online]. Available: http://www.cancer.gov/dictionary?cdrid=44811 [accessed April 21, 2014].

NRC (National Research Council). 2011. Review of the Environmental Protection Agency's Draft IRIS Assessment of Formaldehyde. Washington, DC: National Academies Press.

NRC (National Research Council). 2014. Review of EPA's Integrated Risk Information System (IRIS) Process. Washington, DC: National Academies Press.

NTP (National Toxicology Program). 1999. NTP Technical Report on the Toxicology and Carcinogenesis Studies of Glutaraldehyde (CAS No. 111-30-8) in F344/N Rats and B6C3F1 Mice (Inhalation Studies). NTP TR 490. U.S. Department of Health and Human Services, Public Health Service, National Institute of Health, National Toxicology Program, Research Triangle Park, NC [online]. Available: http://ntp.niehs.nih.gov/ntp/htdocs/LT_rpts/tr490.pdf [accessed July 17, 2013].

NTP (National Toxicology Program). 2010. Report on Carcinogens Background Document for Formaldehyde, January 22, 2010. U.S. Department of Health and Human Services, Public Health Service, National Toxicology Program, Research Triangle Park, NC [online]. Available: http://ntp.niehs.nih.gov/ntp/roc/twelfth/2009/November/Formaldehyde_BD_Final.pdf [accessed July 17, 2013].

NTP (National Toxicology Program). 2011. Formaldehyde. Pp. 195-205 in Report on Carcinogens, 12th Ed. U.S. Department of Health and Human Services, Public Health Service, National Toxicology Program, Research Triangle Park, NC [online]. Available: http://ntp.niehs.nih.gov/ntp/roc/twelfth/profiles/formaldehyde.pdf [accessed July 17, 2013].

NTP (National Toxicology Program). 2013. Draft OHAT Approach for Systematic Review and Evidence Integration for Literature-Based Health Assessments. U.S. Department of Health and Human Services, National Institute of Health, National Institute of Environmental Health Sciences, Division of National Toxicology Program, Office of Hazard Assessment and Translation, February 2013 [online]. Available: http://ntp.niehs.nih.gov/NTP/OHAT/EvaluationProcess/DraftOHATApproach_February2013.pdf [accessed July 17, 2013].

Olsen, J.H., and S. Asnaes. 1986. Formaldehyde and the risk of squamous cell carcinoma of the sinonasal cavities. Br. J. Ind. Med. 43(11):769-774.

Olsen, J.H., S.P. Jensen, M. Hink, K. Faurbo, N.O. Breum, and O.M. Jensen. 1984. Occupational formaldehyde exposure and increased nasal cancer risk in man. Int. J. Cancer 34(5):639-644.

Pinkerton, L.E., M.J. Hein, and L.T. Stayner. 2004. Mortality among a cohort of garment workers exposed to formaldehyde: An update. Occup. Environ. Med. 61(3):193-200.

Roush, G.C., J. Walrath, L.T. Stayner, S.A. Kaplan, J.T. Flannery, and A. Blair. 1987. Nasopharyngeal cancer, sinonasal cancer, and occupations related to formaldehyde: A case-control study. J. Natl. Cancer Inst. 79(6):1221-1224.

Rusch, G.M., J.J. Clary, W.E. Rinehart, and H.F. Bolte. 1983. A 26-week inhalation toxicity study with formaldehyde in the monkey, rat, and hamster. Toxicol. Appl. Pharmacol. 68(3):329-343.

Sellakumar, A.R., C.A. Snyder, J.J. Solomon, and R.E. Albert. 1985. Carcinogenicity of formaldehyde and hydrogen chloride in rats. Toxicol. Appl. Pharmacol. 81(3 Pt 1):401-406.

Soffritti, M., C. Maltoni, F. Maffei, and R. Biagi. 1989. Formaldehyde: An experimental multipotential carcinogen. Toxicol. Ind. Health 5(5):699-730.

Soffritti, M. F. Belpoggi, L. Lambertin, M. Lauriola, M. Padovani, and C. Maltoni. 2002. Results of long-term exposreimental studies on the carcinogeneicity of formaldehyde and acetaldehyde in rats. Ann. N.Y. Acad. Sci. 982:87-105.

Swenberg, J.A., W.D. Kerns, R.I. Mitchell, E.J. Gralla, and K.L. Pavkov. 1980a. Induction of squamous cell carcinomas of the rat nasal cavity by inhalation exposure to formaldehyde vapor. Cancer Res. 40(9):3398-3402.

Swenberg, J., W. Kerns, K. Pavkov, R. Mitchell, and E.J. Gralla. 1980b. Carcinogenicity of formaldehyde vapor: Interim findings in a long-term bioassay of rats and mice. Pp. 283-286 in Mechanisms of Toxicity and Hazard Evaluation: Proceedings of the Second International Congress on Toxicology, July 6-11, 1980, Brussels, Belgium, B. Holmstedt, R. Lauwerys, M. Mercier, and M. Roberfroids, eds. Amsterdam: Elsevier North-Holland Biomedical Press.

Takahashi, M., R. Hasegawa, F. Furukawa, K. Toyoda, H. Sato, and Y. Hayashi. 1986. Effects of ethanol, potassium metabisulfite, formaldehyde and hydrogen peroxide on gastric carcinogenesis in rats after initiation with N-methyl-N'-nitro-N-nitrosoguanidine. Jpn J. Cancer Res. 77(2):118-124.

Til, H.P., R.A. Woutersen, V.J. Feron, V.H. Hollanders, H.E. Falke, and J.J. Clary. 1989. Two-year drinking-water study of formaldehyde in rats. Food Chem. Toxicol. 27(2): 77-87.

Tobe, M., K. Naito, and Y. Kurokawa. 1989. Chronic toxicity study on formaldehyde administered orally to rats. Toxicology 56(1):79-86.

Vaughan, T.L., P.A. Stewart, K. Teschke, C.F. Lynch, G.M. Swanson, J.L. Lyon, and M. Berwick. 2000. Occupational exposure to formaldehyde and wood dust and nasopharyngeal carcinoma. Occup. Environ. Med. 57(6):376-384.

Watanabe, F., T. Matsunaga, T. Soejima, and Y. Iwata. 1954. Study of the carcinogenicity of aldehyde. 1. Experimentally produced rat sarcomas by repeated injections of aqueous solution of formaldehyde [in Japanese]. Gan. 45(2-3):451-452.

West, S., A. Hildesheim, and M. Dosemerci. 1993. Non-viral risk factors for nasopharyngeal carcinoma in the Philippines: Results from a case-control study. Int. J. Cancer 55(5):722-727.

Wilmer, J.W., R.A. Woutersen, L.M. Appelman, W.K. Leeman, and V.J. Feron. 1989. Subchronic (13-week) inhalation toxicity study of formaldehyde in male rats: 8-hour intermittent versus 8-hour continuous exposures. Toxicol. Lett. 47(3):287-293.

Wolf, D.C., K.T. Morgan, E.A. Gross, C. Barrow, O.R. Moss, R.A. James, and J.A. Popp. 1995. Two-year inhalation exposure of female and male B6C3F1 mice and F344 rats to chlorine gas induces lesions confined to the nose. Fundam. Appl. Toxicol. 24(1):111-131.

Woutersen, R.A., L.M. Appelman, J.W. Wilmer, H.E. Falke, and V.J. Feron. 1987. Subchronic (13-week) inhalation toxicity study of formaldehyde in rats. J. Appl. Toxicol. 7(1):43-49.

Woutersen, R.A., A. van Garderen-Hoetmer, J.P. Bruijntjes, A. Zwart, and V.J. Feron. 1989. Nasal tumors in rats after severe injury to the nasal mucosa and prolonged exposure to 10ppm formaldehyde. J. Appl. Toxicol. 9(1):39-46.

Ying, C.J., X.L. Ye, H. Xie, W.S. Yan, M.Y. Zhao, T. Xia, and S.Y. Yin. 1999. Lymphocyte subsets and sister-chromatid exchanges in the students exposed to formaldehyde vapor. Biomed. Environ. Sci. 12(2):88-94.

Zhang, L., C. Steinmaus, D.A. Eastmond, X.K. Xin, and M.T. Smith. 2009. Formaldehyde exposure and leukemia: A new meta-analysis and potential mechanisms. Mutat. Res. 681(2-3):150-168.

Zhang, L., X. Tang, N. Rothman, R. Vermeulen, Z. Ji, M. Shen, C. Qiu, W. Guo, S. Liu, B. Reiss, L.B. Freeman, Y. Ge, A.E. Hubbard, M. Hua, A. Blair, N. Galvan, X. Ruan, B.P. Alter, K.X. Xin, S. Li, L.E. Moore, S. Kim, Y. Xie, R.B. Hayes, M.

Azuma, M. Hauptmann, J. Xiong, P. Stewart, L. Li, S.M. Rappaport, H. Huang, J.F. Fraumeni Jr., M.T. Smith, and Q. Lan. 2010. Occupational exposure to formaldehyde, hematotoxicity and leukemia-specific chromosome changes in cultured myeloid progenitor cells. Cancer Epidemiol. Biomarkers Prev. 9(1):80-88.

3

Independent Assessment of Formaldehyde

The second part of the committee's task was to conduct an independent assessment of formaldehyde. The committee started with its peer review in Chapter 2 and the background document that supports the formaldehyde profile in the 12th RoC. It searched for additional peer-reviewed literature that had been published by November 8, 2013,[1] and incorporated relevant human, experimental animal, and mechanistic studies into the independent assessment. The committee focused its attention on literature that contained primary data, but it also examined published review articles and reviews by other authoritative bodies to ensure that relevant literature was not missed and to ensure that all plausible interpretations of primary data were considered. The committee considered comments and arguments presented to it during its first meeting, comments and documents received from other sources during the study process, and independent literature searches carried out by National Research Council staff (see Appendix D). The goals of the literature searches were to identify relevant literature published around the time of the publication of the background document and later that may have missed inclusion in the 12th RoC and to identify any relevant literature that was published after the release of the 12th RoC. Each search covered the period from January 1, 2009 (the year in which the draft background document for formaldehyde was initially released; Bucher 2013), to November 8, 2013. Databases searched were PubMed, MEDLINE (Ovid), Embase (Ovid), Scopus, and Web of Science. The search strategy for each database is described in Appendix D. After identifying the relevant body of literature up to November 8, 2013, the committee reviewed the primary data and applied the RoC listing criteria to human, experimental animal, and mechanistic studies.

This chapter begins with a section on cancer studies in humans, which is followed by a section on cancer studies in experimental animals. The chapter then reviews toxicokinetic and metabolism literature and studies of mechanisms

---

[1]The cutoff date for the literature search was chosen to allow the committee time to review the literature within the constraints of the project schedule.

of carcinogenesis. It ends with a section that summarizes human, experimental animal, and mechanistic data and provides a conclusion and a listing recommendation for formaldehyde that is based on the listing criteria in the 12th RoC.

The committee's assessment of formaldehyde was guided by the RoC listing criteria, which were first introduced in the present report in Box 1-2. A substance can be classified in the RoC as "reasonably anticipated to be a human carcinogen" if at least one of the following criteria is fulfilled (NTP 2010, p. iv):

- "There is limited evidence of carcinogenicity from studies in humans, which indicates that causal interpretation is credible, but that alternative explanations, such as chance, bias, or confounding factors, could not adequately be excluded."
- "There is sufficient evidence of carcinogenicity from studies in experimental animals, which indicates there is an increased incidence of malignant and/or a combination of malignant and benign tumors (1) in multiple species or at multiple tissue sites, or (2) by multiple routes of exposure, or (3) to an unusual degree with regard to incidence, site, or type of tumor, or age at onset."
- "There is less than sufficient evidence of carcinogenicity in humans or laboratory animals; however, the agent, substance, or mixture belongs to a well-defined, structurally related class of substances whose members are listed in a previous Report on Carcinogens as either known to be a human carcinogen or reasonably anticipated to be a human carcinogen, or there is convincing relevant information that the agent acts through mechanisms indicating it would likely cause cancer in humans."

A substance can be listed as "known to be a human carcinogen" if "there is sufficient evidence of carcinogenicity from studies in humans, which indicates a causal relationship between exposure to the agent, substance, or mixture, and human cancer." The RoC listing criteria are clear about the information needed to fulfill the criteria of sufficient evidence in experimental animals (see the section "Cancer Studies in Experimental Animals"). The type of information needed to meet the RoC listing criteria for limited or sufficient evidence in humans required more interpretation and expert judgment by the committee. To make the committee's methods clear and transparent, the section "Cancer Studies in Humans" begins by describing the committee's methodology for identifying and evaluating epidemiologic evidence and the committee's interpretation and application of the listing criteria.

CANCER STUDIES IN HUMANS

Identification of Informative Epidemiologic Studies

In its independent analysis of formaldehyde exposure and cancers, the committee first considered each of the epidemiologic studies cited in the background document for formaldehyde. As discussed in Chapter 2, the National

Toxicology Program (NTP) did a thorough job of searching the literature for relevant human studies, so the committee used the background document as a starting point for its independent review. Second, the committee examined the results of the independent literature search described in Appendix D (see Box D-1 and Figure D-1). One additional study (Coggon et al. 2014)—an update of Coggon et al. (2003)—was identified after the literature-search cutoff date and was included as part of the committee's independent assessment. Third, the committee examined review articles, meta-analyses, and materials presented during its first meeting and during the study process.

As part of its exclusion criteria (Box D-1), the committee based its assessment on the primary literature. It recognized that quantitative meta-analyses can be informative, but the heterogeneity of exposures in the primary literature on formaldehyde makes it challenging to base any conclusions of causality on resulting summary estimates. The committee agrees with a previous National Research Council report that "meta-analysis can be a valuable method for summarizing evidence but can also be subject to variable interpretations depending on how literature is selected and reviewed and data analyzed" (NRC 2011, p. 112).

Evaluation of Epidemiologic Studies

Several factors were considered in the evaluation of the strength of the epidemiologic literature. The principles of causal association, elaborated by Bradford Hill (Hill 1965), were used as a starting point for the evaluation of informative epidemiologic studies. Of Bradford Hill's original nine criteria, the committee focused on six: strength, consistency, specificity, temporality, biologic gradient, and coherence. On the basis of the RoC listing criteria, plausibility was more relevant to supporting evidence from experimental animal studies and mechanistic data than to the evaluation of the epidemiologic evidence, and analogy was not deemed to be a useful criterion for this topic. Coherence emerged as a particularly important criterion for similarity of findings among multiple study designs and populations (and is also related to consistency). The committee recognizes that the Bradford Hill criteria can be useful guidelines for assessing causal association but agrees with NRC (2014, p. 91) that they "are by no means rigid guides to reaching 'the truth'."

The committee also developed criteria for rating the quality and utility of epidemiologic studies and their exposure assessments, shown in Table 3-1. The development of the exposure-assessment evaluation is presented in detail in Appendix C and summarized in Tables C-1 and C-2. In general, the committee judged a cohort or case–control study to be informative if it was large, had high and varied exposures that were systematically estimated, had reliably assessed cancer end points, and included credible comparison groups. Table 3-2 provides information about all the epidemiologic studies that the committee considered, including a description of the studies, a description of the exposure assessments used in each study, comments on strengths and limitations of the studies, and the committee's determination of study quality (strong, moderately strong, or weak).

TABLE 3-1 Criteria Used to Assess Epidemiologic Studies for Hazard Assessment[a]

Study Quality and Utility Classification	Study Population, Design, Quality of Data, and Analysis	Exposure Assessment[b]
Weak study: limited utility for hazard assessment; inconclusive; uninformative	Modest or small population with few cases. Design limitations, including broad case definition, no duration of exposure, short followup, limited data analysis	Low discrimination between exposed and control categories, qualitative or semiquantitative evaluation, limited evidence of substantial formaldehyde exposure
Moderately strong study: somewhat useful for hazard assessment	Modest-sized population with few cases or a broad case definition; sufficient followup for latency; standard data analysis	Moderate discrimination between high and low exposure categories; substantial fraction of population probably highly exposed; qualitative, semiquantitative, or quantitative evaluation; use of duration of work as a proxy for exposure
Strong study: highly useful for hazard assessment	Large population with many cases, precise case definition, including subcategories; large number of subjects with long-duration exposures; sufficient followup for latency; limited switching among exposure categories; sophisticated data analysis accounting for important potential confounders	High discrimination between high and low exposure categories, substantial fraction of population probably highly exposed, detailed quantitative or highly selective semiquantitative evaluation

[a]The epidemiologic elements in the second column are not required to match with the exposure elements in the third column to define the study quality.
[b]Exposure-assessment levels are based on the data presented in Appendix C and Table C-2.
Source: Committee generated.

The committee's judgment of the strength of a study depended on both the epidemiologic design elements (the second column in Table 3-1) and the exposure-assessment dimensions (the third column in Table 3-1), which are somewhat independent. A strong study might not have a highly developed exposure assessment. For example, several strong case–control studies of licensed embalmers had no exposure assessments, but because the case definition required work as a licensed embalmer and that occupation has well-defined rules for practice (which define the exposure situation), the resulting studies were considered to be strong or moderately strong. A well-designed study with a high-discrimination exposure assessment could be judged to be weak because few of the subjects were exposed to formaldehyde, as was the case, for example, in the textile studies. The overall strength of each study was assessed by considering all of the variables described in Table 3-1.

TABLE 3-2 Description of Epidemiologic Studies Reviewed by the Committee

Reference and Study Population	Study Information[a]	Exposure Assessment[b]	Critique and Conclusions[c]	Study Quality[d]
Andjelkovich et al. 1995 Iron foundry workers from Michigan, USA	Cohort = 8,147 men; outcome: mortality; nasopharyngeal cancer = 1 case, sinonasal cancer = 0 cases, lymphohematopoietic cancer = 15, leukemia = 5 cases; 3,929 workers with potential exposure to formaldehyde for ≥6 months during 1960–1987; 83,064 person-years for exposed and 40,719 person-years for controls; a smoking-history survey was administered via mail.	High-discrimination quantitative exposure assessment; detailed work history available for each study subject; extensive data from industrial-hygienist sampling, technical data from plant, walk-through surveys, and job and task descriptions; information assessed by an industrial hygienist and assigned to high (median 1.5 ppm), medium (median 0.55 ppm), low (median 0.05 ppm), or no formaldehyde-exposure categories; formaldehyde used in core-making operations in 1960–1987; all workers exposed to silica	Followup since first exposure was short (≤19 years), total duration of exposure was short (≤17 years) Although the study had a high-discrimination quantitative exposure assessment and the cohort was of a moderate size, it was probably not large enough to detect risk of rare tumors, such as nasopharyngeal cancer, sinonasal cancer	Moderately strong
Armstrong et al. 2000 General population of Maylasia	Population case–control; outcome: prevalent and incident cases; 282 cases of nasopharyngeal cancers, ≥5 years of residence in study area, and diagnosis in 1987–1992; 282 cases and matched controls identified from health-center records in Kuala Lumpur and Selangor among Malaysian Chinese	Low-discrimination qualitative exposure assessment; exposure information gathered by structured interview to obtain complete dietary, residential, occupational history; exposures classified by broad Malaysian occupational codes, industrial-hygienist professional judgment	Formaldehyde exposure was limited (formaldehyde exposure in only 9.0% of the sample, only eight had accumulated ≥10 years of exposure outside a 10-year latency period); short latency period	Weak
Beane Freeman et al. 2009 NCI study of US chemical industry and plastics workers in 10 plants	Cohort = 25,619; outcome: mortality from lymphohematopoietic malignancy; all lymphohematopoietic types = 319 cases, leukemia = 123 cases, myeloid leukemia = 88 cases; followup period: 1966–2004	High-discrimination exposure assessment; quantitative estimation and job-exposure matrix used, but no measurements after 1980; median exposure intensity was 0.3 ppm (range 0.01–4.3 ppm); median peak exposure was about 2 ppm; about 25% were exposed at >4ppm	Large, well-designed study No evidence of confounding by other exposures Study was able to assess peak exposures	Strong

Study	Description	Exposure assessment	Comments	Rating
Beane Freeman et al. 2013 NCI study of US chemical industry and plastics workers in 10 plants	Cohort = 25,619; outcome: mortality; nasopharyngeal cancer = 10 deaths; sinonasal cancer = 5 deaths; followup period: 1966–2004; update of Hauptman et al. (2004)	High-discrimination exposure assessment; extensive background data and samples; quantitative estimation and job–exposure matrix used on the basis of extensive data, but no measurements after 1980; Beane Freeman et al. (2009) reported the median exposure intensity of 0.3 ppm (range 0.01–4.3 ppm); median peak exposure was about 2 ppm; about 25% were exposed at >4 ppm	Large, well-designed study No evidence of confounding by other exposures Study was able to assess peak exposures	Strong
Bertazzi et al. 1989 Italian resin workers	Cohort = 1,332 men; outcome: mortality; hematologic neoplasms = 7 deaths, lung cancer = 24 deaths, larynx tumors = 6 deaths; followup period: 1959–1986	Moderate-discrimination qualitative exposure assessment; cohort members worked in a department that used formaldehyde; exposure intensity in many locations peaked at >3.0 ppm	Evidence of increasing mortality from hematologic neoplasms with longer latency; highest increase in mortality was in those who were employed during 1965–1969, an early period of high exposure	Moderately strong
Blair et al. 2001 General population in Iowa and Minnesota	Population-based leukemia case-control; outcome: incidence; 513 incident cases; ascertainment period: Iowa 1981–1983, Minnesota 1980–1982	Low-discrimination semiquantitative exposure assessment for formaldehyde; broad job categories and industries; potential formaldehyde exposure was categorized on a 4-point scale; likely high misclassification	There were 513 incident cases, but the study was judged to be weak for assessing formaldehyde because the number of cases with high exposure (n = 3) was small, misclassification likely	Weak
Checkoway et al. 2011 Female textile workers in Shanghai, China	Case–cohort nested within cohort of 267,400 women textile workers; outcome: lung cancer incidence; 628 cases diagnosed in 1989–1998	Low-discrimination qualitative exposure assessment (yes/no) for formaldehyde; detailed job histories and job–exposure matrix used to assign detailed textile-dust and related exposures for all workers for all years; exposure to formaldehyde was uncommon in these workers	Few workers exposed to formaldehyde (2 lung-cancer cases were exposed to formaldehyde)	Weak

(Continued)

TABLE 3-2 Continued

Reference and Study Population	Study Information[a]	Exposure Assessment[b]	Critique and Conclusions[c]	Study Quality[d]
Coggon et al. 2014 Chemical workers in 6 British factories where formaldehyde was produced or used	Cohort = 14,008; outcome: mortality; nasopharyngeal cancer = 1 death, nose and nasal sinus cancer = 2 deaths leukemia = 54 deaths, myeloid leukemia = 36 deaths; followup period: 1941–2012; update of Acheson et al. (1984) and Coggon et al. (2003)	Moderate-discrimination semiquantitative exposure assessment; work histories abstracted from company employment records; jobs were classified into five exposure categories (background, low, moderate, high, or unknown) by industrial-hygiene professional judgment; limited quantitative measurements available after 1970 covering many jobs, quantitative exposure assumed to be the same before 1970 (although anecdotal, the reported exposures were much higher earlier in followup period); "high" exposure category was estimated to be over 2 ppm; no peak exposures identified; authors noted that there was some exposure to paraformaldehyde	Cohort was small and satisfactory for cancers that were more common, but probably too small to detect nasopharyngeal and sinonasal cancers and only had moderate power to detect myeloid leukemia effects Authors reported a concern about the quality of data when they made exposure assignments	Moderately strong
Dell and Teta 1995 Workers employed in a Union Carbide plastics manufacturing plant in New Jersey	Cohort = 5,932; outcome: mortality; nasopharyngeal cancer = 0 deaths, sinonasal cancer = 0 deaths, lymphohematopoietic cancer = 28 deaths, leukemia and aleukemia = 12 deaths; workers employed in 1946–1967; followup through 1988; 5,932 males in the cohort (111 exposed to formaldehyde)	Low-discrimination qualitative exposure assessment; company job histories collected; duration of employment used as a surrogate for cumulative exposure; some analysis of work department made but limited by missing work data	Small study size had little power to detect risk of rare tumors Few workers exposed to formaldehyde Limited exposure information Multiple concomitant exposures (raw materials used in the manufacturing process included asbestos [usually chrysotile], carbon black, epichlorohydrin, polyvinyl chloride, acrylonitrile, styrene, chemical additives [such as plasticizers, emulsifiers, and antioxidants])	Weak

Edling et al. 1987 Workers in abrasive manufacturing in Sweden	Cohort = 521 men; outcome: mortality; nasopharyngeal cancer = 0 deaths; sinonasal cancer = 0 deaths, leukemia = 1 death; men with ≥5 years of employment in 1955–1983; followup period: 1958–1983	Low-discrimination semiquantitative exposure assessment; very limited formaldehyde exposure data from 1970s; two work areas had exposures; blue-collar workers assigned exposures; no data on how many were exposed	Small study size had little power to detect risk of rare tumors Few workers exposed to formaldehyde Limited exposure information	Weak
Hall et al. 1991 UK pathologists	Cohort = 4,512 men; outcome: mortality; nasopharyngeal cancer = 0 cases, sinonasal cancer = 0 cases, leukemia = 4 cases; men identified in 1973 Royal College of Pathologists membership list; followup period: 1974–1987	Low-discrimination qualitative exposure assessment on the basis of job title (formaldehyde exposure was assumed from cadavers); no discussion of exposure conditions was presented	Small study size had little power to detect risk of rare tumor High likelihood of misclassification on exposure to formaldehyde; pathologists have less likelihood of exposures than embalmers	Weak
Hansen and Olsen 1995, 1996 Danish data-linkage study identifying incident cancers in companies in which formaldehyde was used	Cohort = 91,182 men with cancer, 2,041 men with longest work experience of ≥10 years before the date of diagnosis of cancer, 265 companies where formaldehyde was used; outcome: incidence; nasopharyngeal cancer = 4 cases, cancer of the nasal cavity = 13 cases, leukemia = 39 cases; cancer diagnosed in 1970–1984; cases obtained from national cancer registry, linked to national employment data and industry reporting on chemical use	Moderate-discrimination semiquantitative exposure assessment; potentially exposed cases were identified as those with ≥10 years of blue-collar work experience in formaldehyde-using companies; formaldehyde exposures were ranked as low (white-collar jobs) and high (blue-collar jobs) with no wood-dust or high wood-dust exposure; no workplace assessment of exposure conditions or plant size were made, so high potential for misclassification by exposure intensity (for example, a large plant may only have a few workers out of a large workforce who are exposed)	Study limited by lack of data on intensity of exposures and internal plant operations Cohort had no or few cases of some types of cancers, and this limited its utility	Weak

(Continued)

73

TABLE 3-2 Continued

Reference and Study Population	Study Information[a]	Exposure Assessment[b]	Critique and Conclusions[c]	Study Quality[d]
Hauptmann et al. 2009 US funeral directors, embalmers	Nested case–control; outcome: mortality; nasopharyngeal cancer = 4 cases, lymphohematopoitic cancers = 168 cases, myeloid leukemia = 34 cases, brain cancer = 48 cases; those who died in 1960–1986; update of Hayes et al. (1990)	High-discrimination exposure assessment; methods included quantitative reconstruction with statistical modeling, sensitivity analyses; average exposure intensity while embalming was 1.5–1.8 ppm and average peak exposures was 8.1–10.5 ppm depending on case group	No confounding by smoking Strong trend with years in embalming; trends with average and peak exposure	Strong
Hayes et al. 1986 General population in the Netherlands	Case–control; outcome: incidence; histological types of sinonasal cancer = 116 cases; cancer diagnosed in 1978–1981; cases drawn from all six major hospitals for treatment of head and neck tumors	Moderate-discrimination qualitative exposure assessment; work history collected by interview included all jobs held for 6 months or more; all jobs were classified by industrial hygienists according to level and probability of formaldehyde exposure on 10-point scale; agreement between two raters was poor for adjacent scores, and this resulted in high potential for misclassification in adjacent categories, which was rare for high to low or low to high	Study limited by disagreement between exposure assignments of 2 independent raters, but the association of formaldehyde exposure and nasal cancer was similar for each rater For sinonasal cancer, the study suggests an association between formaldehyde and squamous-cell carcinoma, not adenocarcinoma	Moderately strong
Hildesheim et al. 2001 General population in Taiwan	Population case–control; outcome: incidence; nasopharyngeal cancer = 375 cases; newly diagnosed, histologically confirmed nasopharyngeal cancer in people younger than 75 years old who were residents of Taipei City or County for ≥6 months; cases identified at 2 tertiary-care hospitals; population-based controls drawn from national housing registry	Moderate-discrimination semiquantitative exposure assessment; occupational history data obtained by interview; exposures were assigned to broad occupation codes on basis of professional judgment of study industrial hygienist; exposures were classified from 0 (not exposed) to 9 (strong) according to probability, intensity, and duration of formaldehyde exposure; 74 cases exposed to formaldehyde; dietary factors and coexposure to cigarette smoking, wood dust, and solvents were assessed	Considerable overlap in wood dust, formaldehyde exposures; authors were concerned about greater misclassification for formaldehyde than wood-dust assignments >95% of cases were positive for Epstein Barr virus	Moderately strong

Levine et al. 1984 Licensed embalmers in Ontario, Canada	Cohort = 1,477; outcome: mortality; nasopharyngeal cancer = 0 deaths, sinonasal cancer = 0 deaths, larynx = 1 death, lymphohematopoietic cancer = 8 deaths, leukemia = 4 deaths; 34,774 person–years of observation during 1950–1977, 17,589 of which occurred ≥20 years since first licensure	Embalmers have well-defined, high exposures to formaldehyde; embalmer exposure can be sharply discriminated from that of other job groups; job and formaldehyde sources defined by regulations and training	Cohort was small and the study probably had little power to detect risk of rare nasopharyngeal and sinonasal cancers	Moderately strong
Li et al. 2006 Chinese female textile workers in 526 factories in Shanghai	Cohort = 267,400; outcome: incidence; nasopharyngeal cancer = 67 cases, sinonasal cancer = 10 cases; cases identified in 1989–1998; 267,400 female textile workers drawn in 1925–1958	Low-discrimination qualitative exposure assessment for formaldehyde, which was secondary to a primary evaluation of textile production exposures; complete occupational history in textile industry was collected; factory profile form was used by industrial hygienists in Shanghai to record for each factory production processes, types of workshops, and historical measurements of hazardous exposures since establishment of factory	Limited use of formaldehyde in textile operations; very few workers exposed (only 10 cases exposed to formaldehyde and none of the NPC cases were classified as exposed)	Weak
Luce et al. 1993 General population in France	Case–control; outcome: incidence; sinonasal cancer = 207 cases; cases with primary malignancies of the nasal cavity and paranasal sinuses diagnosed in 1986–1988; cases obtained from 27 hospitals, hospital and community controls; analyses performed separately for squamous-cell carcinoma and adenocarcinoma, the two major histologic types	Moderate-discrimination semiquantitative exposure assessment; work history collected by interview; industrial hygienist classified all jobs for probability of exposure (unexposed, possible, probable, definite); 107 cases with exposure to formaldehyde; formaldehyde concentrations in exposed jobs estimated as low (<0.1 ppm), medium (0.1–1.0 ppm), high (>1.0 ppm); authors evaluated coexposures to wood dust	High correlation between wood dust, formaldehyde exposure limited ability to estimate formaldehyde effect separately	Moderately strong

(Continued)

TABLE 3-2 Continued

Reference and Study Population	Study Information[a]	Exposure Assessment[b]	Critique and Conclusions[c]	Study Quality[d]
Luce et al. 2002 General populations in 7 countries	Case–control; outcome: incidence; type of nasopharyngeal cancer: adenocarcinoma = 195 cases, squamous-cell carcinoma = 432 cases; cancer cases diagnosed in 1968–1990; pooled data from 12 case–control studies in seven countries	High-discrimination exposure assessment; uniform methods used in all studies to gather detailed job information; job titles and industries coded uniformly; quantitative exposure data used to construct job–exposure matrix; hygienists assigned probabilities and intensities of formaldehyde exposure; cumulative exposure was principal summary measure of exposure; 192 cases with medium or high exposure to formaldehyde; authors evaluated effects of coexposures to wood dust	Statistical modeling used to evaluate effects of concurrent wood-dust and formaldehyde exposure.	Strong
Luo et al. 2011 General population in 13 US regions covered by SEER registries	Ecologic study; outcome: SEER lung-cancer incidence rates by county; data on age-adjusted lung-cancer incidence rates in 1992–2007; county-level correlation of Toxics Release Inventory data on formaldehyde release with lung-cancer incidence rate from the SEER database	Low-discrimination semiquantitative exposure assessment; county-level quantitative data on industrial release of formaldehyde as proxy for general population exposure in the county	Caution needed in interpreting ecologic associations as causal; high potential for misclassification in large counties	Weak
Mahboubi et al. 2013 General population in Montreal, Canada	Population-based case–control study; outcome: lung-cancer incidence; 1,595 male cases and 465 female cases; interviews conducted in two periods: 1979–1986 and 1996–2002.	Moderate-discrimination semiquantitative exposure assessment; detailed job information gathered by questionnaire; job titles and industries coded uniformly; hygienists assigned confidence, relative concentration, and frequency of formaldehyde exposure; 99 cases with "substantial" exposure to formaldehyde; authors evaluated effects of confounding by smoking and other exposures	Large, well-conducted study; broad job titles limit discrimination Little or no evidence of an association with lung-cancer incidence	Moderately strong

Meyers et al. 2013 US garment-industry workers	Cohort = 11,043; outcome: mortality; nasopharyngeal cancer = 0 deaths, sinonasal cancer = 0 deaths, lymphohematopoietic cancer = 107 deaths, leukemia = 36 deaths, myeloid leukemia = 21 deaths; workers employed for ≥3 months after introduction of formaldehyde-treated fabric into production process (1959 in facilities 1 and 2, 1955 in facility 3); followup through1998; update of Stayner et al. (1985) and Pinkerton et al. (2004)	High-discrimination quantitative exposure assessment; personal exposure samples for formaldehyde from 549 randomly selected employees in five different departments from the 1980s; Pinkerton et al. (2004) reported geometric mean 8-hr TWA of 0.09 ppm–0.20 ppm, overall geometric mean concentration of 0.15 ppm; area monitoring showed that formaldehyde concentrations were essentially constant without substantial peaks or intermittent exposures	Historical data on free formaldehyde in textile fabrics strongly suggest that exposures before 1970 were at least an order of magnitude higher than exposures in the 1980s and later (Elliot et al. 1987) Although the study design was judged to be strong, the cohort was probably not large enough to detect an effect for rare cancers, such as nasopharyngeal cancer, sinonasal cancer	Strong
Olsen and Asnaes 1986 General population in Denmark	Case-control; outcome: incidence; nasopharyngeal cancer = 293 cases, sinonasal cancer = 466 cases; histologically confirmed cancer cases in 1970–1982; male cases and controls selected from Danish Cancer Registry	Moderate-discrimination qualitative exposure assessment; employment histories obtained from national pension, population registries and exposure classified by job description, industry; each job rated by industrial hygienist as unexposed to formaldehyde, probably or certainly exposed, or unknown; wood-products industry is widespread in Denmark	Only small numbers of cases ever exposed to formaldehyde (13 cases of squamous-cell carcinoma; 17 cases of adenocarcinoma ever exposed to formaldehyde); few with formaldehyde exposure and no wood-dust exposure No evidence of confounding by wood dust or smoking	Moderately strong
Ott et al. 1989 Two Union Carbide facilities	Nested case–control; outcome: mortality; lymphohematopoietic cancer = 129 cases, leukemia = 59 cases; cases identified from review of causes of death among males from the Rinsky et al. (1987) cohort who died during 1940–1978; Union Carbide facilities also evaluated by Dell and Teta (1995)	Low-discrimination qualitative exposure assessment for formaldehyde; broad job and plant departments with many exposures and few cases of formaldehyde exposure; formaldehyde exposure was assigned on the basis of work in a department that used formaldehyde	Exposures not localized in production areas, probably resulting in likely broad misclassification Multiple concomitant exposures (raw materials used in the manufacturing process, including asbestos [usually chrysotile], carbon black,	Weak

(Continued)

TABLE 3-2 Continued

Reference and Study Population	Study Information[a]	Exposure Assessment[b]	Critique and Conclusions[c]	Study Quality[d]
			epichlorohydrin, polyvinyl chloride, acrylonitrile, styrene, chemical additives [such as plasticizers, emulsifiers, and antioxidants])	
Partanen et al. 1993 Finnish wood-industry workers	Nested case–control; outcome: incidence; Hodgkin disease = 4, non-Hodgkin lymphoma = 8, leukemia = 12; cancer cases diagnosed in 1957–1982	Moderate-discrimination qualitative exposure assessment; methodology assigned exposure based on personal work histories and a job–exposure matrix that identified formaldehyde exposure; no average exposure intensity was provided	Medium formaldehyde exposures likely, but study limited by small number of cases	Moderately strong
Pesch et al. 2008 German wood industry	Industry-based case–control; outcome: incidence; histologically confirmed sinonasal cancer = 86 cases; recognized occupational disease diagnosed in 1994–2003; cases identified from workers insured by Holz-BG insurance company	Low-discrimination qualitative exposure assessment of formaldehyde; questionnaire collection of occupational history with additional data on wood-related exposures and chemical treatments, including formaldehyde; personal sampling for wood-dust exposure in 1992–2002; expert industrial hygienists estimated wood-dust exposure to identify missing information and trends; crude assessment of formaldehyde exposures (yes/no) with no measurements; 47 cases exposed to formaldehyde (54.6%), an equal fraction of controls	Strong study of wood-dust association with sinonasal cancer, but weak assessment of formaldehyde exposure Substantial exposure misclassification was likely	Weak
Richardson et al. 2008 General population in Germany	Population-based case–control study of non-Hodgkin lymphoma and chronic lymphocytic leukemia; outcome: incidence; non-Hodgkin lymphoma = 858 cases; newly diagnosed cases that occurred in 1986–1998	Low-discrimination semiquantitative exposure assessment; yes/no estimates of formaldehyde exposure derived from job-history data and a job–exposure matrix that used broad job and industry groups	Broad job categories; likely high misclassification of exposure	Weak

Roush et al. 1987 General population in Connecticut	Case–control; outcome: incidence; nasopharyngeal cancer = 173 cases, sinonasal cancer = 198 cases; histologically confirmed cases were from Connecticut Tumor Registry among males who died from any cause in 1935–1975, controls from death certificates	Low-discrimination semiquantitative exposure assessment; occupational histories obtained from death certificates, city directories; exposures were assigned to broad occupation codes on basis of industrial-hygienist professional judgment; high exposure ≥1 ppm	Broad job categories; likely misclassification Risk estimates adjusted for smoking, race, and other risk factors	Weak
Siew et al. 2012 Finnish general population	Cohort = 1.2 million working Finnish men; outcome: incidence; nose = 292 cases, nasal squamous-cell carcinoma = 167 cases, nasopharyngeal cancer = 149 cases; followup period: 1971–1995; data linkage for all men born in 1906–1945 who were employed in 1970	Moderate-discrimination quantitative exposure assessment; occupation in 1970 linked to job-exposure matrix to estimate wood-dust exposure, formaldehyde exposure, coexposures to asbestos and silica; exposure assessment completed by professional industrial hygienists	Few cases with formaldehyde exposure for three of the four types of cancer investigated (17 cases of cancer of the nose, 9 cases of nasal squamous-cell carcinoma, 5 cases of nasopharyngeal cancer, and 1,831 cases of lung cancer with any exposure to formaldehyde) Significant lung cancer–formaldehyde association may have resulted from residual confounding by smoking, wood dust, asbestos, or crystalline silica	Moderately strong
Stellman et al. 1998 American Cancer Society Prevention Study II	Cohort = 362,823 men enrolled in the Cancer Prevention Study-II, 45,399 men employed in a wood-related occupation, reported exposure to wood dust, or both; outcome: cancer mortality; sinonasal cancer = 1 death, nasopharyngeal cancer = 2 deaths, lymphohematopoietic cancer = 122 deaths, non-Hodgkin lymphoma = 51	Low-discrimination qualitative exposure assessment; questionnaire given to self-identified wood workers and others with wood-dust exposure or people who reported exposure to formaldehyde (yes/no), asbestos	High potential for misclassification in self-reporting exposure to formaldehyde	Weak

(Continued)

79

TABLE 3-2 Continued

Reference and Study Population	Study Information[a]	Exposure Assessment[b]	Critique and Conclusions[c]	Study Quality[d]
	deaths, Hodgkin lymphoma = 5 deaths, multiple myeloma = 20 deaths, leukemia = 46 deaths; followup period 1982–1988			
Stern 2003 US tannery workers	Cohort = 9,352 men; outcome: all mortality; nasal = 1 death, leukemia and aleukemia = 16 deaths; included all production workers employed for any length of time at tannery A in 1940–1979 or at tannery B during 1940–1980; followup through 1993; study is an extension of Stern et al. (1987)	Low-discrimination exposure assessment; personnel records were reviewed, subjects were grouped into five departments; semiquantitative potential exposure depended on departments; Stern et al. (1987) reported that ambient formaldehyde was measured in finishing department at time of study and was 0.5–7.0 ppm (mean 2.45 ppm)	Few cases with formaldehyde exposure; standardized mortality ratio for workers in finishing department potentially exposed to formaldehyde	Weak
Stroup et al. 1986 Anatomists living in the United States	Cohort = 2,317 men; outcome: all mortality; buccal cavity and pharyngeal cancer = 1 death, nasal cavity and sinuses = 0 deaths, lymphohematopoietic cancer = 18 deaths, leukemia = 10 deaths, myeloid leukemia = 5 total deaths; men who joined American Association of Anatomists and lived in United States during 1888–1969	Moderate-discrimination exposure assessment; job structure strongly related to exposure; details available for duration of association membership and time period in which anatomists joined the association, which were divided into thirds to provide a crude surrogate of cumulative exposure to formaldehyde; information on research and teaching interests, department affiliations, and membership in other professional associations used to categorize each anatomist as specialist in gross anatomy, microanatomy, both, or neither; on basis of a review of reference materials and on discussions with anatomists who were	Exposure was defined aspect of job and varied according to type of anatomist	Moderately strong

		familiar with laboratory techniques used in past; gross anatomists may have been exposed to formaldehyde more frequently than microanatomists		
Vaughan et al. 1986a General population in western Washington state	Population–based case–control; outcome: incidence reported to cancer registry; all incident cases of pharyngeal cancer (27 cases diagnosed during 1980–1983) and sinonasal cancer (53 cases diagnosed during 1979–1983) in persons between 20–74 years old who resided in the study area	Moderate discrimination semiquantitative exposure assessment; jobs obtained from interview histories were assigned to broad occupation codes; likelihood and intensity of exposure were assigned on basis of industrial-hygienist professional judgment in a 4-category variable; formaldehyde exposure associated with making wood products	Occupational-exposure prevalence was much lower than in West et al. (1993) Only 3.5% of jobs had any formaldehyde exposure (11 cases of nasopharyngeal cancer and 12 cases of sinonasal cancer exposed to formaldehyde above background levels)	Moderately Strong
Vaughan et al. 1986b General population in western Washington state	Population–based case–control; incidence reported to cancer registry; all incident cases of nasopharyngeal cancer (27 cases diagnosed in 1980–1983) and sinonasal cancer (53 cases diagnosed in 1979–1983) in persons between the ages of 20–74 who resided in the study area	Moderate-discrimination semiquantitative exposure assessment; subjects' residential histories, including types of dwelling, were determined from structured telephone interview, which also collected smoking, alcohol, and demographic information; residential history since 1950 included type of dwelling, use of urea-formaldehyde foam insulation, and occurrence of home renovation or new construction with particle board or plywood; information collected on lifetime occupational history to adjust for potential confounding	Although questionnaire data have limited discrimination of past exposures, living in a mobile home has been associated with high formaldehyde exposure in period of about 1950 to middle 1980s	Moderately strong
Vaughan et al. 2000 General population in catchment of 5 US cancer registries	Population-based case–control; outcome: incidence; 196 newly diagnosed nasopharyngeal cancer cases in 1987–1993; cases were identified prospectively	High-discrimination quantitative exposure assessment; detailed job, industry data from structured interviews; each job assessed on basis of industrial-	Large, well-conducted study with high-discrimination exposure assessment; no assessment of peak exposures performed	Strong

(Continued)

TABLE 3-2 Continued

Reference and Study Population	Study Information[a]	Exposure Assessment[b]	Critique and Conclusions[c]	Study Quality[d]
	in five population-based cancer registries in United States; controls identified by random-digit dialing; expanded exposure evaluation relative to Vaughan et al. (1986a)	hygienist professional judgment for probability of exposure and, if exposed, the 8-hr TWA; estimated 8-hr TWA (low <0.10 ppm; moderate ≥0.10, <0.50 ppm; and high ≥0.50 ppm); 13.2% of jobs had ≥10% probability of exposure; coexposure to wood dust was also assessed for each job		
Walrath and Fraumeni 1983 New York state embalmers and funeral directors	Cohort = 1,132 men; outcome: mortality; nasopharyngeal cancer = 0 deaths, sinonasal cancer = 0 deaths, lymphohematopoitic cancer = 25 deaths, leukemia = 12 deaths, myeloid leukemia = 6 deaths, nonwhites had 3 deaths from lymphohematopoitic cancer; persons who died in 1925–1980; 1,132 white, male embalmers and funeral directors licensed in 1902–1980; no duration of employment or length of licensure available; persons who held only funeral director's license were not included	Embalmers make up group that has well-defined high exposures to formaldehyde; tasks and formaldehyde sources are defined by regulations, training; double licensure—embalmer and funeral director—has fewer exposure opportunities	Although the cohort was small, exposures likely to have been substantial with good discrimination and qualitative distinctions between exposed and not exposed

Cohort probably not large enough to detect risk of rare cancers, such as sinonasal cancer, nasopharyngeal cancer | Moderately strong |
| **Walrath and Fraumeni 1984** California state licensed embalmers | Cohort = 1,007 men; cohort: mortality; sinonasal cancer = 0 deaths, lymphohematopoietic cancer = 19 deaths, leukemia = 12 deaths, myeloid leukemia = 6 deaths; men who died in 1925–1980; white male embalmers licensed in 1916–1976; 1,109 deaths; duration of licensure was available but not employment | Embalmers make up group that has well-defined, high exposures to formaldehyde; tasks and formaldehyde sources are defined by regulations, training; length of licensure used as surrogate of length of employment | Although the cohort was small, exposures likely to have been substantial with good discrimination and qualitative distinctions between exposed and not exposed

Cohort probably not large enough to detect risk of rare cancers, such as nasal cancer, nasopharyngeal cancers | Moderately strong |

| West et al. 1993 General population in the Philippines | Population case–control; outcome incidence; nasopharyngeal cancer = 104 cases; followup period: unknown; cases identified at Philippines General Hospital; two types of controls selected: hospital (n = 104) and community controls (n = 101) | Moderate-discrimination semiquantitative exposure assessment; exposure (yes/no) assigned to specific job groups on basis of industrial-hygienist professional judgment; for those exposed, several duration variables were calculated | Association with formaldehyde was stronger for participants who were positive for Epstein Barr virus

No evidence of confounding or effect modification by wood dust or other exposures; estimates adjusted for age, sex, education, ethnicity | Moderately strong |

[a]The study information includes the study type, size of cohort, outcome type, followup period or source of cases and ascertainment period, and prior studies of the same population. The study information also includes the total number of cases by cancer type, which may differ from the number of cases in other tables in Chapter 3 (Tables 3-3–3-7 give the number of cases exposed to formaldehyde).

[b]The exposure-assessment information includes the overall discrimination strength of the study, key data (such as work histories, exposure data and data on jobs, tasks, operations, and key history dates), professional industrial-hygienist data analysis, classification of exposures and metrics used, and data on coexposures. See Table 3-1 and discussion of exposure assessment in Appendix C for descriptions and definitions of terms used in this column.

[c]The committee's critique and conclusions include information on critical study strengths and limitations.

[d]The committee's judgment of the study quality according to the criteria that it developed and presented in Table 3-1.

Abbreviations: ICD, International Classification of Diseases; NCI, National Cancer Institute; ppm, parts per million; SEER, Surveillance, Epidemiology, and End Results program of the National Cancer Institute; TWA, time-weighted average. Source: Committee generated.

Tables 3-3–3-7 present the number of exposed cases for strong and moderately strong studies as a particularly useful indicator of study power. When both disease and exposure are rare, the number of exposed cases will be an important determinant of power (Thomas 2009). The number of exposed cases also has merit because it allows a comparison of size (in the common sense that bigger studies are more powerful) of both case–control and cohort studies. The definition and ascertainment of *exposed* differs among studies and within some studies, so it was sometimes necessary for the committee to make a judgment about which definition to use when choosing the data to present in Tables 3-3–3-7. The reader is referred to the primary literature to view all data and summary measures of exposure reported by specific studies.

As discussed in Chapter 2, particular attention was paid to the choice of summary measures of exposure. Ideally, an epidemiologist chooses the appropriate measure to summarize exposure data on the basis of an understanding or hypothesis about the pharmacokinetics and pharmacodynamics of the exposure-to-dose and dose-to-response processes (Checkoway et al. 2004; Smith and Kriebel 2010). The investigators studying the association between formaldehyde and cancer have little information on which to base that choice. In practice, therefore, it is common and appropriate to test the associations by using several different summary measures, including cumulative exposure, average exposure, duration of exposure, and peak exposure. It is expected that, on average, choosing the wrong metric will result in an underestimation of an association if one exists (Checkoway et al. 2004)—that is, it is not expected that choosing the wrong summary measure of exposure will create evidence of an association where one does not exist except by chance.

Another factor that complicates the assessment of risks by alternative metrics is the imprecision and other limitations of the exposure-intensity data on which the summary measures are based. As discussed above, those data are often only approximations and are likely to have substantial uncertainty. That makes it even more difficult to assert with confidence that one summary measure is more likely than another to be "correct". For those reasons, the committee looked at the measures of association between cancer risk and all the available summary measures presented in each study rather than choosing or preferring one a priori. Furthermore, patterns in disease associations and associated confidence intervals from smaller studies that did not reach traditional significance—that is, a p value less than 0.05 and the exclusion of 1.0 from the 95% confidence interval (CI)—were not discarded in the committee's evaluation of the literature; they were weighed as weaker but still relevant evidence of consistency in the results.

The committee reviewed the available literature on the topic of which exposure metrics are more appropriate for environmental and occupational cancer studies. There is a long history of using cumulative exposure (the product of average intensity and exposure duration) as the summary measure of exposure

(Checkoway et al. 2004). Cumulative exposure tends to be proportional to disease risk and loss of function due to nonmalignant respiratory diseases caused by dusts, such as coal dust, silica, and asbestos. Possibly because of that consistency, cumulative exposure has often been used as the summary measure of exposure for other exposures and other diseases, including cancer. But in the few cases in which data are adequate for examining the relative performance of different exposure metrics, it has been found that cumulative exposure is generally not proportional to cancer risk and should not necessarily be assumed to be the correct summary measure of exposure for cancer risk. Evidence for this finding first came from the studies of Doll and Peto (1978) on smoking and lung cancer, which found that lung cancer risk was not directly proportional to cumulative tobacco exposure (packs/day smoked multiplied by the years of smoking). Cumulative exposure also does not appear to be an appropriate measure for evaluating asbestos exposure and risk of mesothelioma (Peto et al. 1982) and for both asbestos and silica and risk of lung cancer (Zeka et al. 2011). More recently, Richardson (2009) showed that leukemia risk was not proportional to cumulative benzene exposure. In the absence of knowledge about which outcome measure is applicable, the committee concluded that there was no compelling reason to prefer findings for one of the standard exposure metrics mentioned above over another. And, as noted above, the pattern of findings on all available metrics should be evaluated, data permitting.

Consistent with the RoC listing criteria, the committee used its expert scientific judgment to interpret and apply the listing criteria. *Limited evidence* was defined by the committee as evidence from two or more strong or moderately strong studies with varied study designs and populations that suggested an association between exposure to formaldehyde and a specific cancer type, but whose limitations led the committee to conclude that alternative explanations—such as chance, bias, and confounding factors—could not be adequately excluded and that therefore a causal interpretation could not be accepted with confidence. *Sufficient evidence* was defined by the committee as consistent evidence from two or more strong or moderately strong studies with varied study designs and populations that suggested an association between exposure to formaldehyde and a specific cancer type and for which chance, bias, and confounding factors could be ruled out with reasonable confidence because of the study methodologies and the strength of the findings. Consistent with those definitions, the presence of negative findings in other studies, especially weak studies, did not necessarily negate positive findings.

Nasopharyngeal Cancer

The committee reviewed the literature on epidemiologic studies of formaldehyde and nasopharyngeal cancer (see Table 3-3). Vaughan et al. (2000) was a large multicenter case–control study that was conducted in a general

TABLE 3-3 Studies of Nasopharyngeal Cancer and Formaldehyde Exposure

Reference and Study Population	No. NPC Cancer Cases in Exposed	Findings (95% CI)
Beane Freeman et al. 2013 NCI study of US chemical industry and plastics workers in 10 plants	NPC defined by ICD-8 147; number of cases identified from Tables 2–4 in the publication n = 8	OR for highest average intensity of exposure ($\geq$1 ppm) = 11.54 (1.38–96.81) OR for highest peak exposure category ($\geq$4 ppm) = 7.66 (0.94–62.34) and test for trend with increasing peak categories p < 0.005 OR for highest cumulative exposure category ($\geq$5.5 ppm–years) = 2.94 (0.65–13.28)
Hildesheim et al. 2001 General population in Taiwan	Histologically confirmed NPC; number cases identified from Table 2 in the publication Ever exposed to formaldehyde: n = 74 >20 years since first exposure: n = 55	OR for >10 years of exposure = 1.60 (0.91–2.90) OR among formaldehyde-exposed subjects who were positive for Epstein Barr virus = 2.6 (0.87–7.70)
Siew et al. 2012 Finnish general population	Histologically confirmed NPC; number of cases identified from Table 3 in the publication Any exposure to formaldehyde: n = 5	RR (adjusted for wood-dust exposure) for any formaldehyde exposure compared with no formaldehyde exposure = 0.87 (0.34–2.20)
Vaughan et al. 1986a,b General population of western Washington state	NPC defined by ICD code 146-149; number of cases identified from Tables 3 and 5 in Vaughan et al. (1986a) and Table 2 in Vaughan et al. (1986b) n = 11	OR (adjusted for smoking and race) for highest exposure score = 2.1 (0.6–7.8) OR for $\geq$10 years occupational exposure = 1.6 (0.4–5.8) OR for $\geq$10 years of residence in mobile home = 5.5 (1.6–19.4)

Vaughan et al. 2000 General population in catchment of 5 US cancer registries	ICD-O codes used to classify according to three histologic groups of NPC; number of cases identified from Table 2 of the publication Ever exposed: n = 79 Duration >5 years: n = 55	OR for highest cumulative exposure category (>1.10 ppm–years) = 3.0 (1.3–6.6) Positive trend in disease frequency over categories of cumulative exposure (p = 0.033) Wood-dust exposure and smoking had little effect on the relationship with formaldehyde
West et al. 1993 General population in the Philippines	Histologically confirmed NPC; number of cases identified from Table 2 of the publication n = 26 (In some calculations in Table 2 of the publication, n = 27)	OR for ≥25 years since first exposure = 4.0 (1.3–12.3) OR derived from the final model that was adjusted for concurrent effects of education, diesel and dust, smoking, processed meats, fresh fish, mosquito coils, and herbal medicines

Abbreviations: ICD, International Classification of Diseases; NCI, National Cancer Institute; NPC, nasopharyngeal cancer; OR, odds ratio; ppm, parts per million. Source: Committee generated.

population. Incidence data were collected from the National Cancer Institute (NCI) Surveillance, Epidemiology, and End Results Program registries. The study was identified as a strong study (Table 3-1). There were 24 nasopharyngeal-cancer cases in the highest category of cumulative exposure, so this study was one of the largest that the committee reviewed for nasopharyngeal cancer. Its methods included a quantitative exposure assessment with moderate discrimination of who was exposed and the intensity of exposure, and the study was conducted with a well-described expert assessment of formaldehyde exposures classified by self-reported jobs of cases and controls. The estimation of the probability of exposure level or intensity of exposure in each job enabled the investigators to estimate lifetime cumulative exposure of each participant. There was evidence of increasing disease frequency with increasing exposure. The odds ratio (OR) was 3.0 (95% CI 1.3–6.6) for the highest cumulative exposure category (>1.10 ppm-year) compared with nonexposed, and there was a significant trend ($p < 0.001$) in the association between nasopharyngeal cancer and an increasing probability of exposure and duration. Controlling for wood-dust exposure and smoking had little effect on the association. The association appeared to be restricted to squamous-cell carcinoma rather than undifferentiated and nonkeratinizing carcinoma, although this finding is limited by small numbers.

The evidence from the Vaughan et al. (2000) study is supported by several other studies. The National Cancer Institute (NCI) industrial cohort study of mortality is one of the important additional sources of evidence. The committee judged the study to be strong. Since the completion of NTP's assessment of formaldehyde in 2011, the NCI cohort has been updated with 10 additional years of followup: NTP's substance profile for formaldehyde cited Hauptmann et al. (2004), and the update of that study is Beane Freeman et al. (2013). The evidence from the cohort continues to suggest that formaldehyde exposure is associated with an increase in the frequency of nasopharyngeal cancer, although even with the additional followup the numbers of exposed cases are small. There were 10 total deaths from nasopharyngeal cancer (and five total deaths from sinonasal cancer, as discussed below). Although small numbers of cases for rare cancers can be a limitation, even for strong studies, because of the high quality of the quantitative, high-discrimination exposure assessment and the design and conduct of the study, the overall results were considered strong, informative, and continue to be persuasive. In the Beane Freeman et al. (2013) study, there was evidence of increasing mortality with increasing exposure for all three exposure metrics evaluated: average, cumulative, and peak exposure (see Appendix C for discussion of exposure metrics). Compared with low exposure, those in the highest categories of each of those metrics had rate ratios of 11.54 (95% CI 1.38–96.81), 2.94 (95% CI 0.65–13.28), and 7.66 (95% CI 0.94–62.34), respectively. A strength of this study is that there was very little wood-dust exposure (only one case was thought to have had such exposure), so there is little concern that the results were confounded by wood dust (a well-known risk factor for nasopharyngeal cancer).

Several studies were judged to be moderately strong and provided support for the finding of increased nasopharyngeal-cancer risk (Vaughan et al. 1986a,b; West et al. 1993; Hildesheim et al. 2001; Siew et al. 2012). Vaughan et al. (1986a,b) conducted a small population-based case–control study of nasopharyngeal cancer incident cases (n = 27 total cases) that were drawn from 13 counties in western Washington state. Interviews with cases (or next of kin if cases were deceased) and controls provided information on occupation (Vaughan et al. 1986a) and residence (Vaughan et al. 1986b) from which estimates of formaldehyde exposure were developed. There was a weak association between working in a job with formaldehyde exposure and incidence of nasopharyngeal cancer (OR for 10 years or more of exposure compared with none was 1.6, 95% CI 0.4–5.8). There was somewhat stronger evidence of an association between living in a mobile home (a well-documented source of formaldehyde exposure) and incidence of nasopharyngeal cancer (OR for 10 years or more of residence compared with none was 5.5, 95% CI 1.6–19.4) (Vaughan et al. 1986b).

West et al. (1993) conducted a moderately large population-based case–control study of incident cases of nasopharyngeal cancer in the Philippines. The exposure assessment appeared to be a well conducted, semiquantitative assessment with moderate discriminations of exposure and was based on blind expert evaluation of the reported job histories. Several metrics of formaldehyde exposure, particularly in the distant past, were positively associated with nasopharyngeal-cancer incidence. The authors gathered data on several potential confounders, including wood dust, smoking, and dietary factors. In a final model that controlled for confounders, the authors reported that subjects first exposed to formaldehyde 25 years or more prior to diagnosis had an OR of 4.0 compared with never exposed (95% CI 1.3–12.3). Control for smoking and "dust" exposure did not weaken the association.

A somewhat larger population-based case–control study of incident cases with a semiquantitative exposure that had moderate discrimination was conducted in Taiwan by Hildesheim et al. (2001). The exposure assessment was similar to that of West et al. (1993) in that an industrial hygienist reconstructed each subject's occupational history. There was an increased incidence of nasopharyngeal cancer in the longest duration-of-exposure category (OR = 1.60, 95% CI 0.91–2.90), and there was some evidence that the association was stronger in subjects who were seropositive for Epstein Barr virus (OR = 2.6, 95% CI 0.87–7.7).

Siew et al. (2012) used several Finnish national databases to evaluate associations between incidence of sinonasal, nasopharyngeal, and lung cancers and exposures to wood dust and formaldehyde. Cases of those cancers were diagnosed among Finnish men during 1971–1995, and were linked to census data on occupations. A job-exposure matrix was used to estimate wood-dust and formaldehyde exposures for subjects based on their occupations. There were only five nasopharyngeal cancer cases with any formaldehyde exposure and the relative risk (RR) for any formaldehyde exposure compared to no formaldehyde exposure was 0.87. There was a wide confidence interval (95% CI 0.34–2.20).

An industrial cohort study of mortality by Meyers et al. (2013) was judged to be a strong study because it was well-designed with a high-discrimination, quantitative exposure assessment and it included Poisson regression modeling to control for confounding; however, it contributed little information to the evaluation of formaldehyde exposure and nasopharyngeal cancer in that it was not sufficiently large to detect an effect for rare cancers such as nasopharyngeal cancer. There was only a little more than one death expected from nasopharyngeal cancer (n = 1.33), and none were observed.

Several studies that were judged to be moderately strong also contributed little information to the evaluation of nasopharyngeal cancer in that they had a small number of subjects who had nasopharyngeal cancer and were exposed to formaldehyde: Walrath and Fraumeni (1983, 1984), Levine et al. (1984), Stroup et al. (1986), Andjelkovich et al. (1995), and Coggon et al. (2014). Walrath and Fraumeni (1983) reported on proportionate mortality in 1,132 deaths of embalmers in New York. The authors reported that there were no deaths from cancer of the nasopharynx. The authors conducted a similar study of licensed embalmers in California (Walrath and Fraumeni 1984) and again observed no deaths from nasal or nasopharyngeal cancer. The study by Levine et al. (1984) of 1,477 Ontario undertakers with 319 deaths from all causes found one death from cancer of the buccal cavity and pharynx (2.1 expected, standardized mortality ratio [SMR] and CIs not given). The authors did not report whether that death was from nasopharyngeal cancer or a different neoplasm. Stroup et al. (1986) reported a retrospective cohort study of mortality in 2,317 male American anatomists. All or nearly all worked with embalming fluid, which contains formaldehyde and other volatile chemicals. One death from buccal cavity and pharyngeal cancer was observed (6.8 deaths expected, SMR = 0.2, 95% CI 0.0–0.8). The authors did not report whether that death was from nasopharyngeal cancer or a different neoplasm. Andjelkovich et al. (1995) evaluated mortality in a subset of automotive iron-foundry workers in Michigan. The original cohort was 8,147 men, and the subcohort exposed to formaldehyde, 3,929 men. There was one death from nasopharyngeal cancer in the exposed group (no SMR or 95% CI reported). Coggon et al. (2014), an update of the industrial cohort study of mortality by Coggon et al. (2003), reported only one death from nasopharyngeal cancer.

Several studies did not contribute to the committee's assessment of formaldehyde exposure and nasopharyngeal cancer, because the committee judged the studies to be weak and inconclusive (see Tables 3-1 and 3-2). Roush et al. (1987) conducted a population-based case–control study of incidence in 173 men drawn from the Connecticut Cancer Registry who had a history of nasopharyngeal cancer and had died. Occupation was determined from death certificates and city directories. The probable level of formaldehyde exposure was determined from job title, industry, specific employment, and year of employment. For the seven deaths in the highest exposure category—probably exposed to some level of formaldehyde for most of their working life and probably exposed at a high level for 20 years or more prior to death—the OR was 2.3 (95% CI

0.9–6.0; two-sided, p = 0.100), adjusted for age at death, year of death, and availability of occupational information. ORs were given for 14 specific industry categories; none was statistically significant, although numbers were small. Coexposures and residential exposures to formaldehyde were not addressed. Dell and Teta (1995) reported a long-term mortality study of an industrial cohort of workers in a single plastics manufacturing and research and development (R&D) plant in the United States. Of 5,932 male employees, 111 had job assignments that involved formaldehyde. The number of deaths in this small group was not stated, but none was from nasopharyngeal cancer. Hansen and Olsen (1995) investigated cancer incidence in an industrial cohort of men who were employed at 265 companies in Denmark in which formaldehyde exposure was identified. The authors reported standardized proportionate incidence ratios (SPIRs) adjusted for age and calendar period; the comparison group was the Danish population as reported to the Danish Cancer Register. Four cancers of the nasopharynx were reported (3.2 expected, SPIR = 1.3, 05% CI 0.3–3.2). Other coexposures were not reported or adjusted for. Stellman et al. (1998), in an update of the industrial cohort mortality study of the American Cancer Society (ACS) Cancer Prevention Study–II, found one cancer of the nasopharynx in study participants who had an occupational history of exposure to wood dust (OR = 0.44, 95% CI 0.06–3.29) and one in men who had worked in a wood-related occupation (OR 1.44, 95% CI 0.19–10.9). Coexposures were not reported. Armstrong et al. (2000) conducted a large population-based case–control study of nasopharyngeal-cancer incidence (282 cases, all cases were squamous-cell carcinomas) in predominantly Chinese Malaysians. The exposure assessment was qualitative, and the study found no evidence of an association with formaldehyde exposure. Limitations in exposure assessment may contribute to an explanation of the low reported prevalence of formaldehyde exposure (for example, only eight cases reported more than 10 years of exposure and more than 10 years of latency), or formaldehyde exposure may simply have been rare and at low in concentration in the population. In either case, the uninformative finding of this limited study does not weaken the apparent association between formaldehyde exposure and nasopharyngeal cancer. Li et al. (2006) conducted a large industrial cohort study of nasopharyngeal cancer incidence in female textile workers in China that included a low-discrimination, qualitative exposure assessment for formaldehyde (years for ever exposed vs never exposed). The authors noted that there was a potential for formaldehyde exposure to be misclassified. The study had some potential to be informative, but the investigators found few workers who had formaldehyde exposures—10 noncases and no cases were identified as having formaldehyde exposure.

In summary, the committee found that epidemiologic studies provided evidence of a causal association between formaldehyde exposure and nasopharyngeal cancer in humans. Evidence of an association was derived from a strong population-based case–control study (Vaughan et al. 2000), a strong industrial cohort study (Beane Freeman et al. 2013), and several moderately strong population-based case–control studies (Vaughan et al. 1986a,b; West et al. 1993;

Hildesheim et al. 2001; Siew et al. 2012). See Table 3-3 for important key measures of association. The conclusion was based on the strength, consistency, temporality, dose–response relationship, and coherence of the evidence and on the considerations presented in Table 3-1.The most informative epidemiologic studies were ones that were large, that estimated exposure systematically, that had credible comparison groups, and that assessed cancer end points reliably. Not all studies that were judged as strong or moderately strong were informative in the evaluation of the evidence on nasopharyngeal cancer, because of the rarity of tumors at this site and because the studies reported only a few or no deaths from nasopharyngeal cancer. Other studies had sufficient cases but had weak exposure evaluations. The weakest and least informative studies had limited exposure assessments and few or no cases of nasopharyngeal cancer.

Sinonasal Cancer

The committee reviewed the literature on epidemiologic studies of formaldehyde and sinonasal cancer (see Table 3-4). The strongest study was the pooled population-based case–control study by Luce et al. (2002) that assessed incidence data. It provided evidence of an association between formaldehyde exposure and sinonasal cancer. As mentioned in Chapter 2, a pooled study differs from a meta-analysis in that the data from the studies are combined into a single dataset by using the same or similar case definitions and exposure assessments; this is analogous to what is done in a multisite cohort study. The Luce et al. study was particularly valuable because a new exposure assessment was conducted to inform each of the 12 studies that were assembled for the pooled analysis. The exposure assessment was quantitative and had high discriminatory ability; it estimated the level of exposure (average air concentration) and probability of exposure. The exposure data permitted the investigators to analyze risks among categories of cumulative exposure. There was strong evidence of an association between adenocarcinoma and formaldehyde exposure. For example, the OR for sinonasal-cancer incidence was 3.0 (95% CI 1.5–5.7) in men who were in the highest tertile of cumulative formaldehyde exposure compared with no exposure. The comparable OR in women was 6.2 (95% CI 2.0–19.7). The association between formaldehyde and squamous-cell carcinoma was weaker and showed little evidence of a trend. The association between formaldehyde and adenocarcinoma was investigated for possible confounding or effect modification by wood-dust exposure. The researchers used multiple logistic regressions, including analysis of the level of wood-dust exposure as a covariate and stratification on wood-dust exposure, to examine the association between formaldehyde exposure and adenocarcinoma in those who had no wood-dust (or leather-dust) exposure. The results showed only a modest weakening of the formaldehyde risk. In women, the OR for high cumulative exposure fell from 6.2 to

TABLE 3-4 Studies of Sinonasal Cancer and Formaldehyde Exposure

Reference and Study Population	No. SNC Cases in Exposed	Findings (95% CI)
Hayes et al. 1986 General population in the Netherlands	Histologically confirmed ICD-9 160, 160.2–160.5; two raters (A and B) for exposure; number of cases identified from Tables 3 and 4 of the publication	OR for squamous-cell carcinoma cases comparing any vs no formaldehyde exposure = 3.0 (90% CI 1.3–6.4) for rater A, 1.9 (90% CI 1.0–3.6) for rater B
	Any formaldehyde exposure, low wood-dust exposure: rater A, n = 15; rater B, n = 24	OR for squamous-cell carcinoma cases comparing high vs no formaldehyde exposure (with low wood-dust exposure) = 3.1 (90% CI 0.9–10.0) for rater A, 2.4 (90% CI 1.1–5.1) for rater B
	Squamous-cell carcinoma with any formaldehyde exposure, low wood-dust exposure: rater A, n = 12; rater B, n = 19	Rater B assigned proportionally more controls to formaldehyde exposure compared with rater A; rating from both raters showed an increase in OR with increasing formaldehyde assignments
Luce et al. 1993 General population in France	Cancer of nasal cavity and paranasal sinuses ICD-9 160.0, 160.2–160.9; number of cases identified from Table 2 of the publication	OR for adenocarcinoma from possible, probable, or definite formaldehyde exposure and no or low wood-dust exposure = 8.1 (0.9–72.9)
	Adenocarcinoma with probable or definite exposure (male and female): n = 70	
	Squamous-cell carcinoma with probable or definite exposure (male and female): n = 18	
Luce et al. 2002 General populations of 7 countries	Number of cases identified from Table 3 of the publication	OR for adenocarcinoma (adjusted for age and wood- and leather-dust exposure) from high formaldehyde exposure, male = 3.0 (1.5–5.7); female = 6.2 (2.0–19.7)
	Adenocarcinoma cases with medium or high exposure: n = 122 male; 5 female	OR for adenocarcinoma from high formaldehyde exposure and no wood- or leather-dust exposure, male = 1.9 (0.5–6.7); female = 11.1 (3.2–38.0)
	Squamous-cell carcinoma cases with medium or high exposure: n = 70 male; 13 female	OR for squamous carcinoma from high formaldehyde exposure, male = 1.2 (0.8–1.8)

(Continued)

TABLE 3-4 Continued

Reference and Study Population	No. SNC Cases in Exposed	Findings (95% CI)
Olsen and Asnaes 1986 General population in Denmark	Histologically confirmed ICD-7 160.0, 160.2–160.9; number of cases identified from Table 4 of the publication; most formaldehyde exposures occurred in Danish wood-working industry and few formaldehyde cases not exposed to wood dust Ever vs never exposed to formaldehyde: - Squamous-cell carcinoma: n = 13 - Adenocarcinoma: n = 17	Ever vs never exposed to formaldehyde, pooled estimate for formaldehyde exposure adjusted for wood-dust exposure: - Squamous-cell carcinoma of the nasal cavity and sinuses: OR = 2.3 (95% CI 0.9–5.8) - Adenocarcinoma of nasal cavity and sinuses: OR = 2.2 (95% CI 0.7–7.2) ≥10 years since first exposure, pooled estimate for formaldehyde exposure adjusted for wood-dust exposure: - Squamous-cell carcinoma of the nasal cavity and sinuses: OR = 2.4 (0.8–7.4) - Adenocarcinoma of nasal cavity and sinuses: OR = 1.8 (0.5–6.0)
Siew et al. 2012 Finnish general population	Nasal cancer; number of cases identified from Table 3 in the publication Any exposure to formaldehyde: n = 17	RR (adjusted for wood dust) for any formaldehyde exposure compared with no formaldehyde exposure = 1.11 (0.66–1.87)
Vaughan et al. 1986a,b General population in western Washington state	SNC defined by ICD 160: number of cases identified from Tables 3 and 5 in Vaughan (1986a) and Table 2 in Vaughan (1986b) Exposed to formaldehyde above background, n = 12	OR (adjusted for age, sex, smoking, and alcohol) for number of years exposed: 1–9 years = 0.7 (0.3–1.4); ≥10 years = 0.4 (0.1–1.9) OR (adjusted for age, sex, smoking, and alcohol) for cumulative exposure score (all years): 5-19 = 0.5 (0.1–1.6); ≥20 years = 0.3 (0.0–2.3) OR (adjusted for age, sex, smoking, and alcohol) for cumulative exposure score (15-year lag period): 5-19 = 1.0 (0.3–2.9)

Abbreviations: ICD, International Classification of Diseases; OR, odds ratio; ppm, parts per million; RR, relative risk; SNC, sinonasal cancer.
Source: Committee generated.

5.8 (95% CI 1.7–19.4), and males showed a similar reduction. A number of other studies that were judged to be moderately strong contributed to the conclusion that this study was not anomalous. The two key strengths of the Luce et al. (2002) study are the great size and the high-quality exposure assessment; the other studies were smaller and had less adequate exposure assessments. All of the studies have their own limitations, but taken as a whole they provide corroborating evidence.

The moderately strong studies identified by the committee that supported an association between exposure to formaldehyde and sinonasal cancer were Hayes et al. (1986), Olsen and Asnaes (1986), Vaughan et al. (1986 a,b), Luce et al. (1993), and Siew et al. (2012). Hayes et al. (1986) conducted a population-based case–control study of the incidence of histologically confirmed cases of sinonasal cancer in the Netherlands from 1978 to 1981. A low-discrimination, qualitative exposure assessment was conducted independently by two trained hygienists (rater A and rater B) who classified all jobs as to the level (intensity) and probability of formaldehyde (and wood-dust) exposure. The study was large enough to permit separate assessment of risks specifically for cases of squamous-cell carcinoma (there were at least 12 cases with formaldehyde exposure). For all sinonasal cancer combined, the OR was approximately doubled when the exposed were compared with the nonexposed; the CIs excluded 1.0. The authors stratified their analysis by wood-dust exposure (none and low vs high) and found that there were trends of increasing incidence with increasing level of formaldehyde exposure in the no or low wood-dust stratum. That pattern was more evident for squamous-cell carcinomas (there were not enough adenocarcinomas in the group with low wood-dust exposure to permit this analysis). The OR was 3.1 (90% CI 0.9–10.0) for high formaldehyde exposure and low or no wood-dust exposure vs no formaldehyde exposure for rater A and 2.4 (90% CI 1.1–5.1) in the same category for rater B. Rater B assigned proportionally more controls to formaldehyde exposure compared with rater A. The rating from both raters showed an increase in OR with increasing formaldehyde exposure.

Olsen and Asnaes (1986) was an update of Olsen et al. (1984). In the 1986 study, the authors conducted a population-based case–control study of incidence nested in the Danish cancer registry, and they included cancer controls. Denmark has a large wood-working industry, which also includes some formaldehyde exposures. As a result, few cases have formaldehyde exposure without wood-dust exposure. The study had a limited exposure assessment that was based on expert evaluation of job information. The exposure assessment was qualitative and was of moderate discrimination in its assessment in determining whether each subject had certainly or probably been exposed to formaldehyde. The authors investigated separately the association between formaldehyde exposure and incidence of the two main histologic types of nasal and paranasal sinus cancer—squamous-cell carcinoma and adenocarcinoma. When the ever exposed to formaldehyde were compared with the never exposed to formaldehyde, the ORs were very similar for the two subtypes; 2.3 (95% CI 0.9–5.8) for squamous-cell carcinoma and 2.2 (95% CI 0.7–7.2) for adenocarcinoma. Although limited by small numbers, there was evidence of increased incidence of adenocarcinoma from formaldehyde exposure in subjects who were not exposed to wood dust (OR = 7.0, 95% CI 1.1–43.9). When the data were examined for 10 or more years since first exposure, the OR for squamous-cell carcinoma was 2.4 (95% CI 0.8–7.4) and the OR for adenocarcinoma was 1.8 (95% CI 0.5–6.0).

Vaughan et al (1986a) undertook a population-based case–control study in Washington state of 53 incident cases of sinonasal cancer, including 12 in people thought to have had occupational exposure to formaldehyde. The authors found no evidence of increased risk with maximum exposure, number of years exposed, a cumulative exposure score, or the cumulative exposure score with a 15-year lag period. Vaughan et al. (1986b) used the same study group as Vaughan et al. (1986a) to examine the role of residential exposures and sinonasal cancer. Evaluations were reported for people exposed in mobile homes (5 cases, OR = 0.6, 95% CI 0.2–1.7), people living for not more than 10 years in new or renovated housing with particle board or plywood (13 cases, OR = 1.8, 95% CI 0.9–3.8), and people living for 10 years or more in new or renovated housing with particle board or plywood (12 cases, OR = 1.5, 95% CI 0.7–3.2). The authors did not investigate coexposures except for lifetime smoking history and recent consumption of alcoholic beverages.

Luce et al. (1993) conducted a large population-based case–control study (207 cases and 409 controls) of the incidence of sinonasal cancer in France. Histologic data allowed separate investigations of adenocarcinoma and squamous-cell carcinoma. The exposure assessment was semiquantiative with moderate discrimination in that it was based on expert judgment without measurement data for assessment of jobs (which were classified by probability of exposure) and expert assessment of exposure frequency and intensity. The investigators started with a large case series: there were 38 adenocarcinoma cases that had more than 30 years of exposure to formaldehyde. The squamous-cell carcinoma series was somewhat smaller—five in the longest duration category. The study was limited in its ability to discriminate risks associated with potentially confounded wood-dust and formaldehyde exposure, and nearly all cases that had formaldehyde exposure also had probable or definite wood-dust exposure; only four adenocarcinoma cases that had possible, probable, or definite formaldehyde exposure were believed to have had no or low wood-dust exposure (OR = 8.1, 95% CI 0.9–72.9). The authors also reported that the combination of wood dust plus formaldehyde exposure was associated with a higher risk of adenocarcinoma than wood dust alone, although confidence intervals were wide because of the small number of cases.

Siew et al. (2012), the cohort of Finnish men from a national database, was summarized above in the section on nasopharyngeal cancers. There were 17 cases of cancer of the nose and paranasal sinuses in Finnish men identified as having any occupational exposure to formaldehyde. There was a weak association of cancer in those who had any exposure to formaldehyde compared to no exposure to formaldehyde (RR = 1.11, 95% CI 0.66–1.87).

The recently updated NCI industrial cohort study of mortality was judged to be strong, but the number of sinonasal-cancer cases was small (Beane Freeman et al. 2013). There were five deaths from sinonasal cancer in this large cohort (three deaths in the exposed population compared to 3.3 expected deaths). There was no evidence of increased mortality from this cancer, but because of the small numbers of expected deaths from sinonasal cancer, little weight was given to these findings.

Meyers et al. (2013), an update of Pinkerton et al. (2004), was also judged to be a strong industrial cohort study of mortality, but it contributed little information because of its size; there were only 0.95 cases of sinonasal cancer expected and none were observed. The authors investigated mortality in 11,043 workers in three garment plants (Meyers et al. 2013). There were no deaths from sinonasal cancer among in 3,915 deaths reported. Additional details were not provided.

Several studies were judged to be moderately strong, but they contributed little information to the evaluation of sinonasal cancer because few subjects who had sinonasal cancer had been exposed to formaldehyde: Walrath and Fraumeni (1983, 1984), Levine et al. (1984), Stroup et al. (1986), and Coggon et al. (2014). The studies by Walrath and Fraumeni (1983, 1984) were described in the nasopharyngeal-cancer section above; the results of the two studies were not informative for evaluating sinonasal cancer, because no cases were reported. The study by Levine et al. (1984) of a cohort of 1,477 Ontario undertakers with 319 deaths from all causes found no deaths from cancer of the nose, middle ear, or sinuses (0.2 deaths expected, SMR and CIs not given). Stroup et al. (1986) reported a retrospective cohort study of mortality in 2,317 male American anatomists. All or nearly all worked with embalming fluid, which contains formaldehyde and other volatile chemicals. None of the 738 deaths was from cancer of the nasal cavity or sinuses (0.5 deaths expected, SMR = 0, 95% CI 0.0–7.2). Coggon et al. (2014) completed a long-term study of mortality in a cohort of 14,014 men in six British plants where formaldehyde was produced or used. In the group of workers whose jobs that were classified as having potential formaldehyde exposure, there were two deaths from cancer of the nose and nasal sinuses (2.8 deaths expected from US national rates, SMR = 0.71, 95% CI 0.09–2.55). Coexposures were not discussed.

Several studies did not contribute to the committee's assessment of formaldehyde exposure and sinonasal cancer, because the committee judged the studies to be weak and inconclusive (see Tables 3-1 and 3-2). Roush et al. (1987) conducted a population-based case–control study of incident cases in 198 men in the Connecticut Cancer Registry who had a history of sinonasal cancer and died. Occupation was determined from death certificates and city directories. Probable level of formaldehyde exposure was determined from job title, industry, specific employment, and year of employment. The OR for the seven deaths in the highest exposure category was 1.5 (95% CI 0.6–3.9) (adjusted for age at death, year of death, and availability of occupational information). ORs were given for 14 specific industry categories, and none was statistically significant, but the numbers were small. Coexposures and residential exposures to formaldehyde were not addressed.

Dell and Teta (1995) reported a long-term study of mortality in a cohort of industrial workers in a single plastics manufacturing and R&D plant in the United States. Of 5,932 male employees, 111 had job assignments that involved formaldehyde. The number of deaths in this small group was not stated, but none was from sinonasal cancer.

Hansen and Olsen (1995, 1996) conducted a study in a large national cancer cohort of industrial workers and reported SPIRs. The authors obtained government employment data on blue-collar workers employed in Danish industries who were identified as having used formaldehyde and linked those data with cancer-registry data. A national product register was used to identify workers in broad industries in which formaldehyde was used and formaldehyde exposure was likely. The records were used to determine a moderate-discrimination, semiquantitative metric of formaldehyde exposure: duration of work with potential formaldehyde exposure. A similar approach was used to determine wood-dust exposure at the industry level by identifying industrial classification codes that corresponded with jobs that used wood products. Only 13 cases of cancer of the nasal cavity were reported to the national cancer registry (compared with 5.2 deaths expected on the basis of the proportionate distribution of all cancers combined) in men whose longest job was in a company that used formaldehyde. The investigators calculated an SPIR as an estimate of the rate ratio; for nasal cancer, the SPIR was 2.3 (95% CI 1.3–4.0). When the data were limited to blue-collar workers in formaldehyde-using industries in which wood products were not used, the SPIR increased to 3.0 (95% CI 1.4–5.7).

Stellman et al. (1998), in an update of the industrial cohort mortality study of the ACS Cancer Prevention Study-II, found one death from sinonasal cancer in men who had wood-dust exposure and found no evidence of an association with formaldehyde. Stern (2003) completed a study of mortality in an industrial cohort of 9,352 tannery workers in jobs that often included formaldehyde exposure; one death from cancer of the nasal cavity was reported (SMR not given). Pesch et al. (2008) conducted an industry-based case–control study of incident cases of adenocarcinoma of the nasal cavity and paranasal sinuses in the German wood industry (86 male cases, 204 controls). In the group of workers who were exposed to formaldehyde and wood products, eight cases were exposed to formaldehyde before 1985 (OR = 0.46, 95% CI 0.14–1.54), and 39 cases were exposed to formaldehyde in 1985 or later (OR = 0.94, 95% CI 0.47–1.90). Because both cases and controls were exposed to wood dust, a recognized cause of sinonasal cancer, extension to the general population is uncertain.

The committee found that epidemiologic studies provided evidence of a causal association between formaldehyde and sinonasal cancer in humans. Evidence of an association was derived from the strong pooled case–control studies of sinonasal cancer (Luce et al. 2002) and several moderately strong population-based case–control studies (Hayes et al. 1986; Olsen and Asnaes 1986; Vaughan et al. 1986a.b; Luce et al. 1993; Siew et al. 2012). See Table 3-4 for important key measures of association. The conclusion was based on the strength, consistency, temporality, dose–response relationship, and coherence of the evidence and on the considerations presented in Table 3-1.The most informative epidemiologic studies were the ones that were large, that estimated exposure systematically, that had credible comparison groups, and that assessed cancer end points reliably. The studies that did not find associations were usually too small to detect an effect for these rare cancers or used methods of exposure assessment that had little ability to discriminate exposures, and they did not provide convincing evidence that there were sufficient numbers of highly exposed subjects.

Lymphohematopoietic Cancers

The committee reviewed the literature on a potential association between formaldehyde exposure and lymphohematopoietic cancers. This section begins with a discussion of methodologic considerations in exposure assessment in studies of lymphohematopoietic cancers and then discusses in greater detail studies in industrial cohorts and studies in embalmers and others in the funeral trade, anatomists, and pathologists. Data from studies that the committee judged to be strong and moderately strong and informative are presented in Tables 3-5 (industrial workers), 3-6 (funeral workers, embalmers, pathologists, and anatomists), and 3-7 (general population).

Methodologic Considerations in Exposure Assessment in Studies of Lymphohematopoietic Cancers

In the substance profile for formaldehyde, NTP considered the most informative primary studies for the evaluation of lymphohematopoietic cancers to be the study of mortality in the large NCI cohort of formaldehyde-industry workers (Beane Freeman et al. 2009) and the NCI nested case–control mortality study of embalmers and funeral directors, which was based on a cohort of funeral-industry workers (Hauptmann et al. 2009). Those were judged to be the strongest studies because of the high quality of the quantitative exposure assessments, which included assignments of participants into exposure categories with high discrimination.

When large occupational cohorts are used to study relatively rare cancer, subpopulations are drawn from several worksites of varying size to obtain sufficient cases. Although the worksites have exposure to formaldehyde as a common feature, they can have large differences in exposure conditions even if the job titles and types of operations are the same (see Appendix C for a more detailed discussion). Beane Freeman et al. (2009) conducted a comprehensive exposure assessment, which increases confidence that valid exposure–response trends can be derived from the diverse industries and exposure conditions.

Both the formaldehyde-industry (Beane Freeman et al. 2009) and funeral-industry (Hauptmann et al. 2009) cohorts included extensive separate evaluations of occupational exposures, their determinants, and modeling approaches to reconstructing unmeasured historical exposures.[2] The exposure studies of the formaldehyde-industry cohort were reported by Blair et al. (1986, 1990) and Hauptmann et al. (2004). The exposure studies of the funeral-industry cohort were reported by Stewart et al. (1992). The committee recognized that those additional exposure studies were keys to the strength of the epidemiologic studies. Because Beane Freeman et al. (2009) and Hauptmann et al. (2009) were critical for the formaldehyde assessment of lymphohematopoietic cancers, this section elaborates on their approaches.

The exposure assessments for the formaldehyde-industry and funeral-industry cohorts were designed to determine exposures associated with job titles and worksites listed in the work histories of the study subjects so that exposures and subjects could be linked. Historical changes in job activities and in the formaldehyde industry produced substantial differences in temporal profiles of exposure. Industrial exposures have declined considerably since the early 1970s as a result of process changes and engineering controls of process emissions. The exposures in the Beane Freeman et al. (2009) study changed (more in some jobs than in others), and the data suggest that exposures in the 1960s were much higher than those after 1970 (Blair et al. 1986, 1990). Embalming-fluid emissions of formaldehyde have probably changed little, but local exhaust ventilation was added in some funeral homes and was estimated to have reduced exposure by 50–90% (Stewart et al. 1992).

Exposures in the industrial and embalming settings were described by time-weighted averages (TWAs) and short-term measurements. The short-term measurements were used to capture brief (15 minutes) intense exposures called peaks. Although peaks are part of the distribution of short-duration concentrations that contribute to the longer TWA measurements, they might not correlate well with the overall average (Blair and Stewart 1990), as was seen in the Beane Freeman et al. study (2009). Blair and Stewart (1990) also noted that exposure metrics can differ among manufacturing plants because in some plants everyone is exposed but in others only half the workforce is in areas with exposure or because similar work areas had lower exposures.

As explained in Appendix C, the summary measures of exposure (which are also called exposure metrics or dose metrics) used in epidemiologic studies are weighting schemes applied to summarize the complex temporal profiles of personal exposure histories. In that application, they are analogous to the concept of dose applied in toxicologic studies, but there is no universal dose metric that applies to all toxic responses, including carcinogenesis. Some dose metrics are not appropriate for the underlying biology, and when an inappropriate metric is used, a weaker or no dose–response relationship will usually be observed

---

[2]Appendix C provides a general summary of exposure assessments, the rationale for estimating exposures on the basis of physical principles, and a description of methods for measuring airborne formaldehyde exposures.

(Blair and Stewart 1990; Smith and Kriebel 2010). Although cumulative exposure is the most common dose metric for chronic, minimally reversible disease processes, it is probably not the optimal dose measure for studying cancer (Smith and Kriebel 2010), as noted above. A fundamental feature of cumulative exposure is that it gives equal weight to long, low-intensity exposures and short, high-intensity exposures, which may not be biologically appropriate for cancer biology. A lag time until effects are observed may also be included in the exposure metric to account for an induction period between the first exposure to formaldehyde and the diagnosis of cancer. That period includes any delay from first exposure to the exposure that initiated the cancer, the time from initiation through the biologic events that led to malignant change, and the time required for that change to produce signs or symptoms that result in diagnosis. Those steps are commonly thought to require at least 10 years for solid cancers in adults, perhaps less for leukemia and lymphomas.

Epidemiologic models that use exposure metrics for peak exposures hypothesize an underlying nonlinear damage process in which exposures at low concentrations have little or no effect and exposures at high concentrations produce disproportionate effects. That might indicate a threshold process, or some protective process might be overwhelmed or a damaging secondary process might occur. When the mode of action is unknown, it is common for epidemiologists to try several exposure metrics, such as cumulative exposure and peak exposure that have different biologic implications (Blair and Stewart 1990).

The mechanistic process associated with the cumulative exposure and peak exposure metrics appear to be different, and conceptually the metrics should be useful for obtaining insight about the possible mechanism of the effects. Unfortunately, the precision of estimated metric values is often limited by sparse historic data and the cost of making measurements, variation of exposure between subjects, process and material variation in the industrial operations, and business and economic variations in the demand for a product. If the precision is too limited, it may not be possible to determine which metric is the strongest. Data quality and extrapolation approaches may favor one dose metric over another. Thus, as discussed above, it is common for epidemiologists to calculate several different exposure metrics, such as cumulative exposure, average exposure, and the occurrence or frequency of peaks. When data and resources are limited, epidemiologists often use simpler metrics, such as years of work in a job, categories of ever exposed vs never exposed on the basis of job title or work location, or sometimes even 'ever having worked in an exposed industry'.

In addition to the NCI formaldehyde-industry study (Beane Freeman et al. 2009) and the NCI nested case–control study (Hauptmann et al. 2009), Meyers et al. (2013), an update of Pinkerton et al. (2004), was considered to have strong methods (Table 3-2). The study investigated mortality in an industrial cohort of garment workers. The authors relied on earlier studies of the same sites by Stayner et al. (1985, 1988), Acheson et al. (1984), and Gardner et al. (1993).

Semiquantitative exposure estimates were developed on the basis of small numbers of measurements, job activities, and reports of sensory irritation in jobs or work locations.

There were also several moderately strong studies of limited utility in industrial workers (Bertazzi et al. 1989; Partanen et al. 1993; Andjelkovich et al. 1995; Coggon et al. 2014) and embalmers, anatomists, or pathologists (Walrath and Fraumeni 1983, 1984; Levine et al. 1984; Stroup et al. 1986). Those had smaller populations and less discriminating exposure assessments and as a result contributed less to the evidence of an association between formaldehyde and lymphohematopoietic cancers than did the strong studies. Most of the smaller studies used job information alone to define those who were "exposed"—an approach that has little ability to discriminate among people with varied levels of exposures. Duration of exposure obtained from occupational histories was used as a semiquantitative exposure metric, but again, duration alone does not discriminate among exposures that have different intensities.

Population-based case–control studies have the most serious problem of exposure misclassification because they draw from the broad mixture of personal and industrial activities throughout the population in a wide area. For example, the broad job categories of "mortician" and "undertaker" include embalmers (the most highly exposed) but also include a number of less exposed occupations. People in some of those other occupations may occasionally do embalming, but less frequently, and embalming is not one of their main job activities. The categories also include funeral directors, who usually do not embalm. And differences are related to the size of funeral homes' businesses. Use of narrow, well defined, specific job titles, such as a focus on embalmers, can greatly reduce misclassification even without specific measurements.

Studies of Industrial Cohorts Exposed to Formaldehyde

Table 3-5 provides the studies of industrial cohorts exposed to formaldehyde that the committee judged to be strong or moderately strong. As already stated, the NCI industrial-worker cohort mortality study is large, well conducted, and informed by a quantitative, high-discrimination exposure assessment (Beane Freeman et al. 2009). The investigators collected mortality data on workers employed in US chemical factories that used formaldehyde during 1966–2004. The study was the largest in terms of numbers of exposed cancer cases—there were 286 hematologic-malignancy cases, including 116 leukemia cases, and 44 of the leukemia cases were classified as myeloid leukemia. Exposure levels varied widely over time and among plants; the estimated overall median daily exposure was 0.3 ppm. The manufacturing plants produced a various of products, including formaldehyde (plants 2, 7, and 10), formaldehyde resins and molding compounds (plants 1, 2, and 7–10), molded plastic products (plants 8 and 9), photographic film (plants 4 and 5), decorative laminates (plant 6), and plywood (plant 3) (Blair et al. 1990).

TABLE 3-5 Lymphohematopoietic Cancers: Industrial Workers

Reference and Study Population	No. Cancer Cases in Exposed				Findings (95% CI)			
	All Lymphohematopoietic Cancer	Leukemia	Myeloid Leukemia	All Lymphohematopoietic Cancer	Leukemia	Myeloid Leukemias
Andjelkovich et al. 1995 US iron-foundry workers (Number of cases from Table 3 of the publication)	7	2	—	SMR = 0.59 (0.23–1.21)	SMR = 0.43 (0.05–1.57)	—
Beane Freeman et al. 2009 NCI study in US chemical workers (Number of cases from Table 1 of the publication)	286	116	44	peak >4 ppm: RR = 1.37 (1.03–1.81), trend with increasing peak exposure	peak >4 ppm: RR = 1.42 (0.92–2.18), trend with increasing peak exposure	peak >4 ppm: RR = 1.78 (0.87–3.64) highest peak category before 1994: RR = 2.79 (1.08–7.21), p trend = 0.02
Bertazzi et al. 1989 Italian resin workers (Number of cases from Table 3 of the publication)	7	—	—	SMR = 7/3.9 = 1.8 (0.72–3.70)	—	—
Coggon et al. 2014 UK chemical workers (Number of cases from Table 6 of the publication)	—	18	9	—	high exposure ≥1 year: OR = 0.59 (0.23–1.50)	high exposure: OR = 1.26 (0.39–4.08)

103

(Continued)

TABLE 3-5 Continued

Reference and Study Population	No. Cancer Cases in Exposed			Findings (95% CI)		
	All Lymphohematopoietic Cancer	Leukemia	Myeloid Leukemia	All Lymphohematopoietic Cancer	Leukemia	Myeloid Leukemias
Meyers et al. 2013 Update of Pinkerton et al. (2004) US garment workers (Number of cases from Table 2 of the publication)	107	36	21	SMR = 1.11 (0.91–1.34)	≥10 years of exposure and ≥20 years since first exposure: SMR = 1.74 (1.10–2.60)	≥10 years of exposure and ≥20 years since first exposure: SMR = 1.90 (0.91–3.50) 16–19 years exposure vs none: SRR = 6.42 (1.40–32.30); test for trend with increasing duration: p = 0.01
Partanen et al. 1993 Finnish wood-industry workers (Number of cases from Tables 1 and 3 of the publication)	7	2	—	OR = 2.49 (0.81–7.59)	OR = 1.40 (0.25–7.91)	—

Abbreviations: CI, confidence interval; OR, odds ratio; RR, relative risk; SMR, standardized mortality ratio; SRR, standardized rate ratio.
Source: Committee generated.

That complexity might have introduced problems of noncomparability among the plants, but a thorough reconstruction of historical formaldehyde average and peak exposures was conducted consistently for all sites until 1980. Good-quality historical data on potential confounders were also assembled from plant records and interviews of long-term employees. Because it pooled data from many plants, the study was powerful enough to detect effects that would not be measurable in plant-by-plant analyses. The formaldehyde exposure assessment was conducted only for jobs held until 1980. Thus, there is likely to have been more error in the exposure assignments in the later time period; in the primary analyses, exposure after 1980 was assumed to be zero. Two sensitivity analyses were conducted to evaluate the effect of that assumption on the results.

About one-fourth of the NCI industrial-worker cohort was estimated to have experienced peak exposures of at least 4.0 ppm (Beane Freeman et al. 2009). A 1999 Agency for Toxic Substances and Disease Registry literature review found that the threshold for mild to moderate human eye, nose, and throat irritation by formaldehyde ranged from 0.4 to 3 ppm in 17 laboratory studies (ATSDR 1999). Thus, the highest peak exposure category (greater than 4 ppm) was above the irritation threshold, and at this level about 50–100% of subjects would have experienced an irritation response.

There was evidence of increased risk of myeloid leukemia with increasing formaldehyde exposure (Beane Freeman et al. 2009). The evidence was strongest when the peak-exposure metric was used, weaker when average exposure was used, and very weak when the effect of cumulative exposure was assessed. In the primary analysis (which assumed zero exposure for all jobs after 1980), the RR of myeloid leukemia increased with increasing exposure. Compared with those who had peak exposures less than 2.0 ppm, the RR in those who had peak exposures from 2.0–4.0 ppm was 1.30 (95% CI 0.58–2.92) and in those who had peak exposures of at least 4.0 ppm, 1.78 (95% CI 0.87–3.64). The data also show the expected pattern wherein the RRs for the highest peak category compared with the lowest peak category increased as the tumor category was narrowed—the RR of all lymphohematopoietic cancers was less than that of all leukemias grouped, and the RR of all leukemias grouped was less than that of myeloid leukemias grouped. The associations were weaker when average exposure was used as the summary measure of exposure than when peak exposure was used, but the trends were similar. A modest increase in RRs was observed among categories of increasing average exposure. The RR increased from the group of all lymphohematopoietic cancers to the grouping of all leukemias, and the RR increased further from the grouping of all leukemias to the grouping of myeloid leukemia.

Beane Freeman et al. (2009) investigated the sensitivity of their results to the assumption of zero exposure after 1980 by censoring all persons who were still exposed in 1979 (this resulted in a loss of about 5% of the person–time of followup). The resulting effect estimates were stronger for both peak and average exposure metrics. For example, the RR for the highest peak exposure category increased from 1.79 (cited above) to 2.64 (95% CI 1.12–6.20), and the trend among categories was also stronger (p = 0.03). The authors reported that there were stronger associations with exposures in the distant past, which may be explained either by higher air concentrations or by a relatively short latency for formaldehyde-induced leukemia. There was evidence to support the former explanation; exposures in the plants were much higher before 1970 than in later years when exposure controls were instituted (Stewart et al. 1986). The possibility of a relatively short latency (compared with that of solid tumors) is supported by two studies of the association between benzene and leukemia (Silver et al. 2002; Glass et al. 2004). In both cohorts, the RR of leukemia after benzene exposure decreased with increasing follow up, and the authors proposed that this is likely due to a relatively short latency for the effects caused by benzene.

Beane Freeman et al. (2009) reported that for the period up to 1994, the RR for the highest peak-exposure category compared with the lowest was 2.79 (95% CI 1.08–7.21), and there was evidence of an increasing trend among categories (p = 0.02). It is not clear why Beane Freeman et al. (2009) found an association with peak exposure and not with cumulative exposure. The committee noted that there were only 10 cases of myeloid leukemia in the highest cumulative exposure category, which was defined as at least 5.5 ppm-years. That is not very many cases and not a very high level of exposure. As a result, this finding is not strong evidence against an association between formaldehyde and myeloid leukemia.

As noted earlier in this chapter, the alternative exposure metrics of peak, average, and cumulative exposure are expected to be proportional to the incidence of a disease as related to different biologic mechanisms or pathways. A complicating factor that must also be considered is the effect of exposure assessment errors on the resulting summary measures. However, it cannot be predicted with any confidence which exposure metric would be expected to be closer to the "truth" in the investigation of formaldehyde and cancer. Therefore, the committee assessed peak, average, and cumulative exposure with equal weight on its overall evaluation. More precise studies in the future may be able to resolve this issue.

Hodgkin lymphoma was strongly associated with peak exposure (RR = 3.96, 95% CI 1.31–12.02) when the subgroups with the highest and lowest peak exposure were compared. A positive association with multiple myeloma was also observed when the highest and lowest peak-exposure subgroups were compared (RR = 2.04, 95% 1.01–4.12). For both outcomes, there was evidence of a trend of increasing mortality with increasing peak exposure. The findings on Hodgkin lymphoma and multiple myeloma are potentially important for further investigation, but the committee did not find additional evidence of these associations in other studies.

An important strength of the NCI industrial-cohort study was its ability to investigate possible confounding by other chemical exposures (antioxidants, asbestos, benzene, carbon black, dyes and pigments, hexamethylenetetramine, melamine, phenol, plasticizers, urea, and wood dust); none was found. Beane Freeman et al. (2009) specifically investigated a potential confounding effect of benzene by excluding all workers who were known to have been exposed to benzene, and the results were not changed. Plant heterogeneity was investigated and found not to be an important factor in the results. There were some limitations. Despite the size of the study, the numbers of deaths in some categories of rare neoplasms were still small, and this limited the power to detect associations in the smallest subgroups. The magnitude of the exposure–response associations changed over time, and it is not possible without strong a priori assumptions to distinguish alternative explanations, such as disease latency, changes in exposures associated with changes in industrial operations and engineering controls, or time-dependent measurement uncertainties.

The committee concluded that although those limitations exist, the study was of high quality. The careful and clearly documented design and analysis reduced the likelihood that the results could be explained by bias. As noted, the authors investigated important sources of confounding and found no important evidence of confounding that might seriously undermine their results. Chance is an unlikely explanation given the consistent patterns of increased RR among exposure categories and tumor categories noted above. Thus, the committee determined that the findings are relevant to evaluating an association between formaldehyde exposure and myeloid leukemia.

Additional evidence of an association between formaldehyde exposure and lymphohematopoietic cancers in workers who were exposed during industrial operations was found in the National Institute for Occupational Safety and Health (NIOSH) study of garment workers. Meyers et al. (2013) updated earlier reports by Stayner et al. (1988) and Pinkerton et al. (2004) on mortality in a cohort of 11,043 industrial workers who were exposed to formaldehyde in three garment-manufacturing plants. The cohort was considerably smaller than the NCI formaldehyde-industry cohort (21 myeloid-leukemia deaths compared with 44 in the NCI cohort). The study methods included a high-discrimination, quantitative exposure assessment for current exposures that was performed during the early 1980s, which was an important strength of the study, but it did not cover the full period of exposures. The investigators did not attempt to estimate earlier exposures. The only known source of formaldehyde exposure was off-gassing from treated fabrics (which were produced elsewhere), so the amount of free formaldehyde in the fabric was a primary determinant of the workroom exposure (Elliot et al. 1987). Before 1970, the free-formaldehyde content of the fabric was estimated to be over 4,000 ppm; by 1980, the fabric concentrations had been reduced to 100–200 ppm. The air concentration measured in the workrooms in 1984 (geometric mean exposure, 0.15 ppm) was a result of off-gassing of the 100–200 ppm in the fabric. The ratio of fabric content to air content was about 1,000:1. Assuming that the ratio is fairly constant, fabric that contained 4,000 ppm probably produced an air concentration of about 4 ppm before 1970. However, the investigators did not make use of that simple estimate of earlier exposure; they merely noted that air exposure was likely to have been higher before 1970. Goldstein (1973) reported that industry efforts to reduce formaldehyde levels in work rooms by reducing the amount of resin in the fabric resulted in a decreased from 10 ppm in 1968 to 2 ppm in 1973. Formaldehyde air concentrations were found to be similar between plants and across departments within the same plant. TWA concentrations were reported in a fairly narrow range (0.09–0.20 ppm), and there was little evidence that short-term peaks exceeded the mean. Given the relatively homogenous exposure scenario, it was reasonable to use all employed workers as the exposed group and to compare their mortality with that in the general population. They used years of work from the workers' company job histories to approximate cumulative exposure and implicitly assumed that each year had roughly the same intensity of exposure, so the cumulative exposures of the workers who entered the cohort before 1970 were substantially underestimated.

The committee considered Meyers et al. (2013) to be a strong study for the evaluation of formaldehyde and myeloid leukemia. The study found evidence of an association with myeloid leukemia. The committee reviewed the evidence from both Meyers et al. (2013) and Pinkerton et al. (2004) together because the only important difference between them was that the former had 10 more years of followup (through 2008 instead of 1998). As noted earlier, some evidence in the literature on benzene and leukemia suggests risks decrease with increasing followup (Silver et al. 2002; Glass et al. 2004), and this pattern was observed in the two analyses of the NIOSH garment workers cohort. With followup through 1998, the SMR for all leukemia in those who had an exposure duration of 10 years or more and whose time since first exposure was 20 years or more was 1.92 (95% CI 1.08–3.17); with 10 additional years of followup, the SMR decreased to 1.74 (95% CI 1.10–2.60). For myeloid leukemia, the SMR for the same exposure definition as above with followup through 1998 was 2.55 (95% 1.10–5.03); with followup through 2008, it was 1.90 (95% CI 0.91–3.50). There was little evidence of increased mortality from lymphocytic leukemia in either reports of the NIOSH garment-workers cohort (Pinkerton et al. 2004; Meyers et al. 2013).

The Meyers et al. (2013) report included additional Poisson regression modeling of the data on all leukemia and myeloid leukemia. Those analyses enabled better control of confounding and a more thorough investigation of alternative exposure metrics than were available in Pinkerton et al. (2004). There was a strong positive trend in mortality with increasing duration of formaldehyde exposure (p = 0.01). The standardized rate ratio for 16–19 years of exposure was 6.42 (95% CI 1.40–32.20), although the rate ratio dropped in the longest duration category, at least 19 years. Again, that decrease may reflect the pattern of decreasing risk with extended followup.

The garment workers' coexposures were generally different (lint particles and cleaning-solvent vapors) from those of the NCI formaldehyde-industry cohort, and this reduced the likelihood that an unmeasured confounder would explain both associations. No other potentially carcinogenic exposures were identified in the plants. As noted above, the exposure assessment had some important limitations. However, the committee agreed with the authors that it is reasonable to assume relatively constant exposure intensity throughout the period of employment. On balance, the committee concluded that the finding of an association between formaldehyde exposure and an association with myeloid leukemia was unlikely to have been explained by an unknown bias or confounder, and chance was an unlikely explanation given the pattern of statistically significant findings.

Coggon et al. (2014), an industrial cohort study of mortality in UK chemical workers, was judged to be moderately strong. The publication was an update of Coggon et al. (2003) and included 12 additional years of followup and more than 2,000 additional deaths. The earlier study included very few leukemia deaths and did not provide data specifically on myeloid leukemia. In some respects, Coggon et al. (2014) is similar to the NCI formaldehyde-industry study, but it is smaller and provides less information on its exposure assessment. The 2014 update included substantially fewer exposed myeloid-leukemia deaths; for example, there were nine deaths with "high" exposure in Coggon et al. (2014) and 19 deaths in Beane Freeman et al. (2009) with peaks greater than or equal to 4.0 ppm. Coggon et al. (2014) benefited from a semiquantitative exposure assessment that provided moderate discrimination among jobs with varied exposure intensities. Work histories were abstracted from employment records. Each job was classified into one of five exposure categories—background, low, moderate, high, or unknown—by an industrial hygienist who used professional judgment. Quantitative environmental measurements were available after 1970 that covered many jobs, but the authors judged the data insufficient to estimate cumulative exposure or other formal metrics. Exposures were assumed to be the same before 1970 (although anecdotally reported exposures were much higher earlier in the followup period). Peak exposures were not evaluated, nor were temporal trends evaluated or estimated. The authors reported that "each job title [within a factory] was assigned to the same exposure category across all time periods" (Coggon et al. 2014). More than 95% of subjects were exposed before the middle 1980s, and less than 5% of the cohort was still working after the middle1980s. The authors extended the followup of a previously reported cohort of 14,014 men (Acheson et al. 1984; Gardner et al. 1993) who had worked in six plants where formaldehyde was made or used. Mortality was compared with national rates in England and Wales and, in some cases, local rates. Coggon et al. (2014) mention several coexposures, but they do not provide details or report adjusted rates. In the most detailed exposure–response analysis, a nested case–control study, ORs for myeloid leukemia were estimated for four categories of exposure intensity and for a duration 5 years before disease onset. No analysis by duration, cumulative exposure, or other standard continuous exposure metric was presented. CIs for the effect estimates were wide and included the null value. An effect of the size observed in the NCI cohort would probably not have been detectable, so although the results were not inconsistent with those of Beane Freeman et al. (2009), Hauptmann et al. (2009), and Pinkerton et al. (2004), the committee determined that, on balance, the study was generally inconclusive.

The committee judged three additional studies of small industrial cohorts that evaluated formaldehyde and lymphohematopoietic cancers to be moderately strong (Bertazzi et al. 1989; Partanen et al. 1993; Andjelkovich et al. 1995). Each was based on only a handful of cases. Two of the three yielded some evidence of an association with lymphohematopoietic cancers (Bertazzi et al. 1989 and Partanen et al. 1993). Bertazzi et al. (1989) reported on cancer mortality in an industrial cohort of 1,330 male workers who produced formaldehyde resins, including 219 for whom specific work histories could not be determined. Among the 179 deaths, there were seven from lymphohematopoietic cancer; 3.9 deaths were expected from national rates and 4.9 deaths expected from local rates, but regardless of which standard was used, the observed excess could have been due to chance. For the entire category of lymphohematopoietic cancers, the authors reported an SMR of 5.35 (95% CI 1.56–14.63) in plastic-resin workers who had formaldehyde exposures during 1965–1969, a period that had no exposure controls and therefore likely high exposure. Formaldehyde exposures before 1975 were often greater than 2.4 ppm (3.0 mg/m^3). Duration of work in the plant was often short. There was no discussion of possible coexposures. The seven cases of lymphohematopoietic cancer were not further categorized, so no analyses for leukemia was possible. Partanen et al. (1993) conducted a small industrial nested case–control study of the incidence of lymphoma and leukemia in Finnish wood-industry workers who were exposed to formaldehyde. There were only two exposed leukemia cases (type unspecified) with an adjusted OR for formaldehyde exposure of 1.40 (95% CI 0.25-7.91). The Andjelkovich et al. (1995) industrial cohort study of foundry workers examined mortality in 3,929 men who had potential exposure to formaldehyde for at least 6 months during their work in a single automotive iron foundry. Comparisons were with the US population and with workers in the plant who were not exposed to formaldehyde. There were two deaths from leukemia (type not specified) in exposed workers and three deaths from leukemia in unexposed workers. The study was too small to be informative.

Studies of Embalmers and Others in the Funeral Trade, Anatomists, and Pathologists

Table 3-6 summarizes the studies that the committee judged to be strong or moderately strong that investigated embalmers and others in the funeral trade, anatomists, and pathologists. NCI assembled and followed a cohort of inactive or deceased embalmers and funeral directors (Hauptmann et al. 2009). The study is particularly useful for evaluating the association between formaldehyde exposure and cancer because of the likelihood of high exposures and a high-quality exposure assessment that was conducted by Stewart et al. (1992) and extended by Hauptmann et al. (2009). The authors conducted a nested case–control analysis of data on the cohort, using mortality as the outcome measure. The case

TABLE 3-6 Lymphohematopoietic Cancers: Funeral Workers, Embalmers, Pathologists, and Anatomists

Reference and Study Population	No. Cancer Cases in Exposed — All Lymphohematopoietic Cancer	Leukemia	Myeloid Leukemia	Key Measures of Association (95% CI) — All Lymphohematopoietic Cancer	Leukemia	Myeloid Leukemia
Hauptmann et al. 2009 US funeral directors, embalmers (Number of cases identified from Tables 1 and 2 of the publication)	168	44 (lymphohematopoietic malignancy of nonlymphoid origin)	33	Ever embalm: OR = 1.4 (0.8–2.6)	Ever embalm: OR = 3.0 (1.0–9.5)	Ever embalm: OR = 11.2 (1.3–95.6) Highest level of all exposure metrics had p<0.05
Levine et al. 1984 ON provincial licensed embalmers (Number of cases identified from Table 1 of the publication)	8	4	—	O/E = 1.2 (0.53–2.43)	O/E = 1.6 (0.44–4.10)	—
Stroup et al. 1986 US anatomists (Number of cases identified from Table 3 of the publication)	18	10	3	SMR = 1.2 (0.7–2.0)	SMR = 1.5 (0.7–2.7)	SMR = 8.8 (1.8–25.5)
Walrath and Fraumeni 1983 NY state-licensed embalmers	25	12	6	PMR = 1.2 (0.79–1.79)	PMR = 1.4 (0.73–2.47)	PMR = 1.5 (0.54–3.19)

(Continued)

111

TABLE 3-6 Continued

Reference and Study Population	No. Cancer Cases in Exposed			Key Measures of Association (95% CI)		
	All Lymphohematopoietic Cancer	Leukemia	Myeloid Leukemia	All Lymphohematopoietic Cancer	Leukemia	Myeloid Leukemia
(Number of lymphohemtopoietic and leukemia cases identified from Table 3 of the publication; number of cases of myeloid leukemia noted on page 408 of the publication)						
Walrath and Fraumeni 1984 CA state-licensed embalmers (Number of lymphohemtopoietic and leukemia cases identified from Table 3 of the publication; number of cases of myeloid leukemia noted on page 4640 of the publication)	19	12	6	PMR = 1.2 (0.73–1.90)	PMR = 1.8 (0.90–3.04) PMR for ≥20 years of licensure = 2.2	PMR = 1.5 (0.55–3.26)

Abbreviations: CI, confidence interval; O/E, observed/expected; OR, odds ratio; PMR, proportionate mortality ratio; SMR, standardized mortality ratio. Source: Committee generated.

subjects were 6,808 embalmers and funeral directors who died during January 1, 1960–January 1, 1986, and deaths were included if they had an underlying or contributory cause identified as lymphohematopoietic cancers of lymphoid origin (99 cases) or nonlymphoid origin (48 cases). Myeloid leukemia (34 cases) was analyzed as a separate subgroup. The control subjects were identified randomly from people in the funeral industry who died of other causes, excluding cancers of the buccal cavity and pharynx, of the respiratory system, and of the eye, brain, or other parts of the nervous system. A quantitative exposure assessment was conducted by using information on workplaces and job tasks drawn from interviews with former co-workers and next of kin (Hauptmann et al. 2009) and a NIOSH air-monitoring study (Stewart et al. 1992). All subjects had interview job histories that indicated funeral home or not, embalming or not, and funeral-home ventilation characteristics, which were the predominant factors that affected exposures. The authors found that the average exposure intensity during embalming was 1.7 ppm.

The study group was relatively large: there were 34 myeloid-leukemia deaths in the latest followup (33 had "ever embalmed") (Hauptmann et al. 2009), nearly as many as the 44 in the NCI formaldehyde-industry cohort (Beane Freeman et al. 2009). The findings of Hauptmann et al. (2009) point strongly toward an association between formaldehyde exposure and myeloid leukemia, although measures of associations were stronger in the broad category of all lymphohematopoietic cancers and all leukemias. The simplest exposure metric—distinguishing ever vs never embalming—was moderately associated with increased mortality from all lymphohematopoietic cancers (OR = 1.4, 95% CI 0.8–2.6), more strongly associated with mortality from all leukemias (OR = 3.0, 95% CI 1.0–9.5), and strongly associated with increased myeloid leukemia mortality (OR = 11.2, 95% CI 1.3–95.6). There was a trend of increasing mortality with increasing duration of embalming (p = 0. 02), rising to OR = 13.6 (95% CI 1.6–119.7) when the group that had more than 34 years of embalming was compared with the group that had never embalmed. There was also a clear trend (p = 0.04) with increasing peak exposure, which is a metric similar to the one that Beane Freeman et al. (2009) found to be associated with myeloid leukemia in the different setting of the NCI industrial-cohort workers. In the highest peak-exposure category (greater than 9.3 ppm), the OR was 13.0 (95% CI 1.4–116.9) compared with no exposure. Another similarity to the findings of Beane Freeman et al. (2009) was that there was not a clear trend of increasing mortality with increasing cumulative exposure (p = 0.19).

Hauptmann et al. (2009) found no evidence of an association between formaldehyde exposure and leukemia of lymphoid origin. The specificity within the broader grouping increased the committee's confidence that the results were not likely to be due to an unknown bias. A striking finding of the study was that of the 34 myeloid-leukemia cases, only one did not ever embalm. The ratio of 33:1 contrasts with the ever: never embalming ratio of roughly 4:1 in controls (the exact numbers were 210:55). The 4:1 ratio is a simple way to see the associations noted above by using different exposure metrics, but it created a methodologic limitation for the authors in that the unexposed reference group only had one case. That limitation reduced the precision of the OR reported above. To investigate the effect, the authors repeated the analyses with an enlarged "unexposed" group, which included those who reported fewer than 500 embalming procedures in their career. As expected, the measures of association in the redefined reference group were lower than those reported above, but the patterns were very similar. For example, the OR for those who reported more than 34 years of embalming was 3.9 (95% CI 1.2–12.5) compared with the OR of 13.6 reported above.

Strengths of Hauptmann et al. (2009) were that high exposures were readily identified and there were good supporting data on the range for exposure assignments (Stewart et al. 1992). The model used by the authors explained a high percentage of variability of exposure measurements (74%) (Hauptmann et al. 2009). Errors in quantification would probably not affect the relative ranking of individual exposure histories, especially in the high-exposure category. There was no evidence of confounding by smoking, and few additional chemicals that might confound the association with formaldehyde were involved. In addition, the authors did not adjust for possible changes in work or employer; this could lead to overestimates or underestimates of exposure. The total duration of embalming work was estimated for all subjects, but some exposure information was missing. Exposures from large spills were important for peaks but infrequent and generally not recorded. The authors also noted that "there was a considerable amount of missing data that required imputation for analyses" (Hauptmann et al. 2009, p. 1697). However, sensitivity analyses suggested that the key findings were unaffected by the absence of some data points.

On balance, the committee concluded that Hauptmann et al. (2009) was a strong study. The committee did not identify any important biases that might have explained the key finding of an association between formaldehyde and myeloid leukemia. The authors persuasively demonstrated that confounding was an unlikely explanation. In addition, the clear pattern of associations with multiple increasing exposure metrics and after several sensitivity analyses makes it unlikely that chance could have explained the findings.

Several small studies of embalmers (Walrath and Fraumeni 1983, 1984; Levine et al. 1984) and anatomists (Stroup et al. 1986) in the 1980s provided supporting evidence and were judged to be moderately strong. Each study had only a handful of leukemia deaths and inadequate exposure assessment that was based on the high likelihood of job exposure to formaldehyde and documentation of years of work. Three of the four studies found a pattern of increasing mortality from leukemia in general and from myeloid leukemia specifically, although few were statistically significant; Walrath and Fraumeni (1983, 1984) and Stroup et al. (1986) provided data on myeloid leukemia as the cause of death.

Walrath and Fraumeni reported proportionate mortality ratios (PMRs) and proportionate cancer mortality ratios (PMCRs) in a cohort of embalmers in New York State (1983) and California (1984). The PMRs for all leukemias combined were 1.2 (based on 12 deaths) and 1.8 (based on 12 deaths) in New York and California, respectively. Confidence intervals were not given in the publication, but they were calculated by the committee (see Table 3-6). There was a small excess in PMRs among workers who had less than 20 years of experience and a statistically significant excess in those who had more than 20 years. The authors noted that embalming fluid contains potentially carcinogenic substances other than formaldehyde.

Levine et al. (1984) studied mortality in a cohort of 1,477 licensed undertakers in Ontario and found four deaths from leukemia, not further specified (2.5 deaths expected, SMR not given).The authors also presented a brief analysis of mortality in formaldehyde-exposed men in eight plants and cohorts of pathologists and anatomists; when the results were combined with their own study of undertakers, 53 leukemia deaths were observed and 44 deaths expected. The publication does not provide additional details.

Stroup et al. (1986) reported a retrospective cohort mortality study of 2,317 anatomists, who are exposed to a wide array of solvents, stains, and preservatives, including formaldehyde. The authors found 10 deaths from leukemia (6.8 deaths expected, SMR = 1.5, 95% CI 0.7–2.7). Information on potential confounders and biases was not presented, but the authors suggested that low SMRs for smoking-related cancers and cirrhosis of the liver suggested that cohort members used cigarettes and alcohol less than the general population.

Other Studies Potentially Relevant to Formaldehyde and Lymphohematopoietic Hematologic Cancers

The committee reviewed all other studies in the background document for formaldehyde for evidence bearing on the question of the carcinogenicity of formaldehyde. Studies that were reviewed were judged to be weak and contributed no informative evidence to this review of lymphohematopoietic cancers were those by Edling et al. (1987), Ott et al. (1989), Hall et al. (1991), Dell and Teta (1995), and Stern (2003). Each was small with a low-discrimination exposure assessment that did not permit reliable estimation of an association between formaldehyde exposure and any of the types of cancers of interest. The study by Edling et al. (1987) was a cohort study of mortality that focused on abrasives and leather tanneries, respectively, and formaldehyde constituted a secondary exposure. Hall et al. (1991) updated a study of mortality in a cohort of 4,512 British pathologists (Harrington and Oakes 1984) and found four deaths from leukemia (2.63 deaths expected, SMR = 1.52, 95% CI 0.41–3.89). Followup was nearly complete. Coexposures were not discussed. Dell and Teta (1995) and Ott et al. (1989) studied the same large chemical plants that manufacture a variety of chemicals; few people were exposed to formaldehyde, and the broad job titles limited the specificity of exposure assignments. Dell and Teta (1995) reported on mortality in a cohort of 5,932 male employees in a plastics manufacturing and R&D facility in New Jersey. SMRs for leukemia and aleukemia were 0.98 in hourly employees (12 deaths observed, 12.31 deaths expected, 95% CI 0.50–1.70) and 1.98 in salaried employees (11 deaths observed, 5.56 expected, 95%

CI 0.99–3.54) in salary employees. Numerous possible coexposures were mentioned by the authors. The text reports eight leukemia deaths (three expected) in the R&D workers, but does not include details. Dell and Teta (1995) provided no data on lymphohematopoietic cancers and formaldehyde. Ott et al. (1989), building on a cohort mortality study by Rinsky et al. (1987), conducted a nested case–control study of mortality in male workers in two chemical-manufacturing facilities and an R&D center in New Jersey. The four causes of death that they studied included nonlymphocytic leukemia. Controls were group-matched on decade of first employment and survival. Exposure was assessed on the basis of departmental usage; coexposures were numerous. There were two cases of nonlymphocytic leukemia (2.6 expected, SMR not given). The Stern (2003) study followed mortality in a cohort of workers in two leather tanneries. It had no formal assessment of formaldehyde exposure, and workers were exposed to many toxic agents, including possible carcinogens. Comparisons were with both US and state rates. There were 16 deaths from leukemia and aleukemia (22 deaths expected according to US rates, SMR = 0.72, 95% CI 0.41–1.18). Results in the two tanneries were similar, as were SMRs based on state rates. There was little evidence of a trend with years of employment. The study did not break down leukemia mortality to permit assessment of the myeloid subgroup.

The committee also identified several studies based on general-population registries or surveys that it judged to be weak and that contributed little or no evidence to this review of lymphohematopoietic cancers. Blair et al. (2001) was a population-based case–control study of 513 incident cases and 1,087 matched controls. It focused on agricultural risk factors in leukemia cases drawn from cancer registries in Iowa and Minnesota. The authors investigated workers who had job-related chemical exposures. In those whose work histories suggested low or high formaldehyde exposure, the ORs for chronic myeloid leukemia were 1.3 in the low-exposure category (7 cases, 95% CI 0.6–3.1) and 2.9 in the high-exposure category (1 case, 95% CI 0.3–24.5). Coexposures were numerous. Richardson et al. (2008) conducted a population-based case–control study of non-Hodgkin lymphoma and chronic lymphocytic leukemia incidence in Germany. Semiquantitative estimates of formaldehyde exposure derived from job-history data, and a job–exposure matrix were weakly positively associated with non-Hodgkin lymphoma and chronic lymphocytic leukemia, but confidence intervals were wide and included the null. The study did not address myeloid leukemia.

Hansen and Olsen (1995), which was a Danish cancer incidence study, was described earlier because it found an increased incidence of sinonasal cancer in formaldehyde-exposed workers. The authors reported an SPIR for leukemia in men who worked in 265 factories that imported or manufactured formaldehyde. They found 39 leukemia deaths (47.0 deaths expected, SPIR = 0.8, 95% CI 0.6–1.6). Coexposures were not investigated. The exposure definition used in the study (being a blue-collar worker in a company that was registered with the government as a user of formaldehyde) probably led to substantial misclassification with the likely consequence of underestimation of true risks. Another limitation of the study was that it did not report results separately for leukemia types. For all leukemia types combined, the study did not find evidence of an increased incidence in formaldehyde-exposed workers, although the confidence interval was wide (SPIR = 1.0, 95% CI = 0.6–1.4).

Stellman et al. (1998) analyzed cancer mortality in members of the ACS Cancer Prevention Study II, a very large prospective industrial cohort study. Mortality was examined after 6 years in 45,399 men who had reported being employed in wood industries or occupationally exposed to wood dust and 362,823 who did not report such exposures. Thirty-two leukemia cases were observed in those who reported wood-dust exposure (SMR = 0.90, 95% CI 0.63–1.30), and 14 were observed in the partially overlapping group in wood-related occupations (SMR 1.08, 95% CI 0.6–1.85). The exposure assessment for formaldehyde was by self-report alone, which is likely to be of poorer quality than an expert review and job–exposure matrix. Furthermore, the authors did not report results for subtypes of leukemia. As a result, this study was judged to be of little utility for the committee's assessment.

Summary of Evidence on Lymphohematopoietic Cancers

In summary, the committee concluded that the epidemiologic studies provided evidence of a causal association between formaldehyde and myeloid leukemia in humans. Evidence of an association was derived from two strong industrial cohorts (Beane Freeman et al. 2009; Myers et al. 2013), one strong cohort of embalmers (Hauptmann et al. 2009), and several moderately strong cohorts from the chemical industry (Coggon et al. 2014) and the funeral trade (Walrath and Fraumeni 1983, 1984; Stroup et al. 1986). See Tables 3-5 and 3-6 and Figures 3-1 and 3-2 for key measures of association supporting this conclusion. The conclusion was based on the strength, consistency, temporality, dose–response relationships, and coherence of the evidence according to the quality criteria presented in Table 3-1.

To present data from the studies, it was necessary to choose a particular exposure definition; however, it is important to note that, in its evaluation of the body of evidence, the committee did not choose a single exposure metric a priori for analysis. Instead, it looked at the full set of exposure metrics and their associations with disease.

Figure 3-1 emphasizes a pattern noted earlier—that is, in the studies that were large enough and detailed enough to present associations between formaldehyde and the "nested" case definitions of all types of lymphohematopoietic cancers, all leukemias, and myeloid leukemia, the measures of association tended to increase as the definition was narrowed (the data points for the nested sets of case definitions are linked by a solid line in Figure 3-1).The figure also illustrates that the stronger and larger studies generally reported stronger associations with formaldehyde and were more likely to present confidence bounds for their

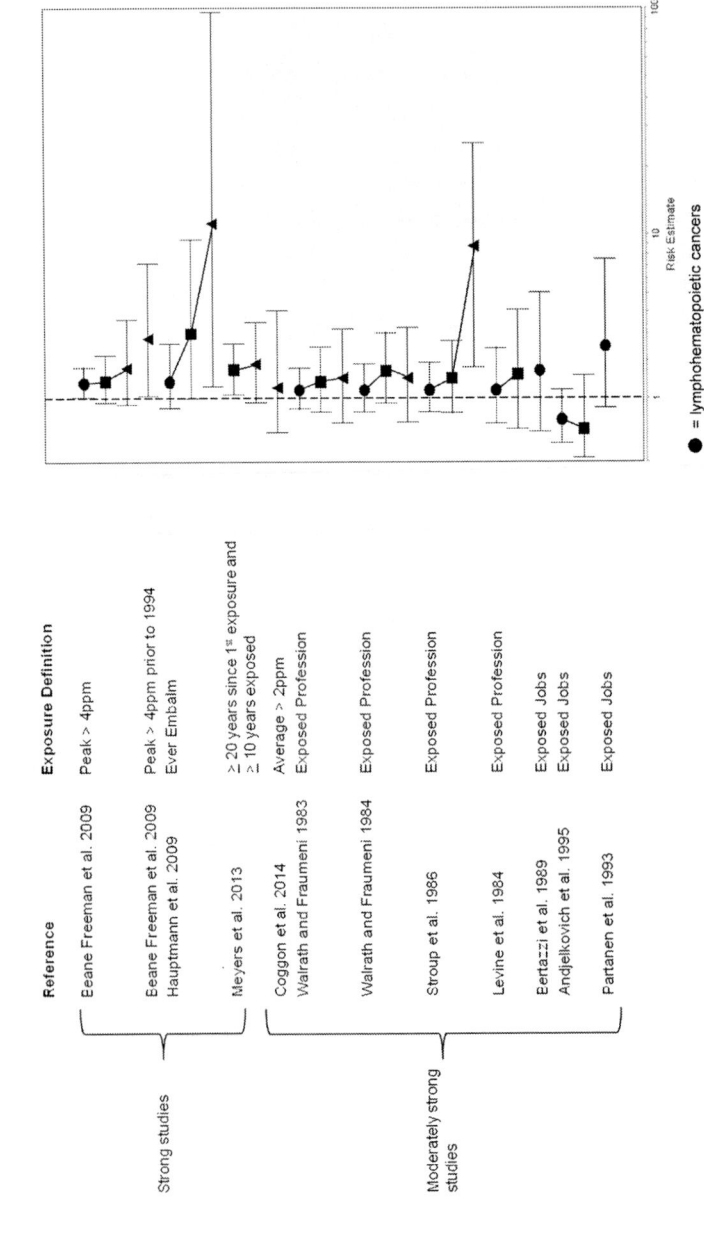

FIGURE 3-1 Summary of strong and moderately strong studies of formaldehyde and lymphohematopoietic cancers. Note: Data points connected by a line indicate results from the same study according to the same exposure metrics but for different tumor sites.

Independent Assessment of Formaldehyde 119

effect estimates that excluded the null. Measures of association between formaldehyde exposure and myeloid leukemia are represented in Figure 3-2 for all studies that reported this association. There is a pattern of positive findings from studies that were judged to be large and strong studies.

Low-precision studies, such as those with a small cohort, only a few cases, or limited exposure assessments, may provide some useful data on risk estimates if several studies were performed. When several small populations are studied using a good design, the measures of association would not be expected to be the same. They would have a distribution that would cluster around the overall risk value for the population; some estimates would be above that value and some would be below that value. If the risk estimates for formaldehyde exposure and myeloid leukemia showed a distribution that was shifted above 1.0 so that few studies showed RRs below 1.0, that pattern of results suggests that there may be a causal relationship between exposure and disease risk. The closer the risk values cluster around 1.0 (some above and some below), the less likely it is that a relationship exists. In Figures 3-1 and 3-2, nearly all RRs are above 1.0, which suggests that a relationship exists. That argument does not imply that all studies are equal. Strong studies make more precise estimates of the RR and are more useful in assessing factors that may affect the RR compared with weaker studies. Strong studies should not produce large RRs when the relationship is weak or absent unless there is a bias in the data.

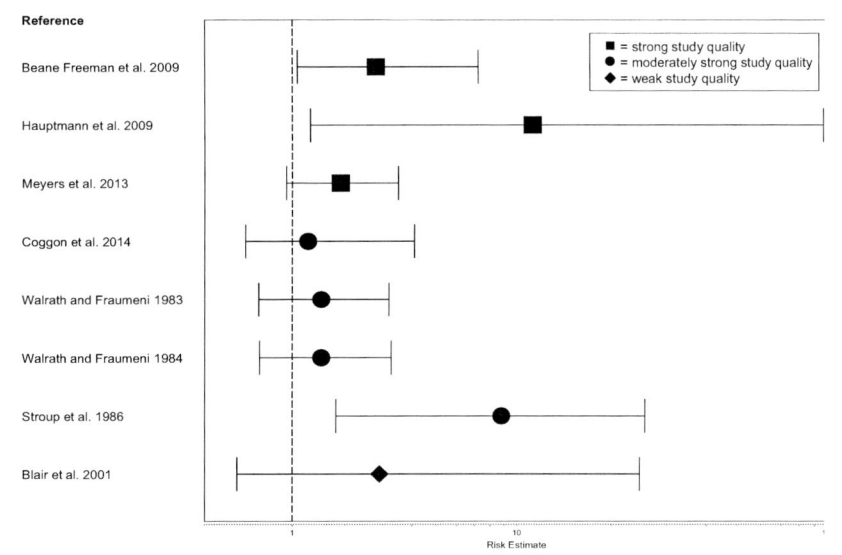

FIGURE 3-2 Summary of key findings from all studies that reported associations between formaldehyde and myeloid leukemia.

As noted above, the informative epidemiologic studies were the ones that were large, that estimated exposure systematically, that had credible comparison groups, and that assessed cancer end points reliably. Studies that did not find associations between exposure and myeloid leukemia were usually too small to detect an effect, did not break out results for myeloid leukemia, or used methods of exposure assessment that resulted in exposure misclassification. A single, large, high-quality study (Beane Freeman et al. 2009) found evidence of increased risk of Hodgkin lymphoma and multiple myeloma in those who had a history of high peak exposures. Those findings do not appear to be supported by other epidemiologic evidence and, in the committee's view, constitute insufficient evidence of effects.

Cancer at Other Sites

The committee conducted a literature search (see Appendix D) to identify studies that examined associations between formaldehyde and cancers at other sites (Table 3-7). Four studies were identified that reported measures of association between formaldehyde and lung cancer. Two of the studies were judged to be moderately strong (Siew et al. 2012; Mahboubi et al. 2013) and two studies were judged to be weak (Checkoway et al. 2011; Luo et al. 2011).

TABLE 3-7 Other Cancer Sites

Reference and Study Population	No. Lung Cancer Cases in Exposed	Findings (95% CI)
Checkoway et al. 2011 Female textile workers in Shanghai, China	Number of cases identified from Table 3 of the publication Cases with ≥10 years of formaldehyde exposure: n = 2	Hazard ratio for ≥10 years formaldehyde exposure = 2.1 (0.4–11.0)
Luo et al. 2011 General population in 13 US regions covered by SEER registries	Not relevant; unit of analysis was county	RR for counties with any formaldehyde release vs none = 1.14 (1.05–1.24)
Mahboubi et al. 2013 General population in Montreal, Canada	Number of cases identified from Table 3 of the publication Cases with "substantial" exposure: n = 99	OR for pooled population comparing substantial with no exposure = 0.88 (0.63–1.24) No evidence of trend with duration, time since first exposure
Siew et al. 2012 Finnish general population	Number of cases identified from Table 3 of the publication Cases with any formaldehyde exposure: n = 1,831	RR for any formaldehyde exposure = 1.18 (1.12–1.25)

Abbreviations: CI, confidence interval; OR, odds ratio; RR, relative risk; SEER, Surveillance, Epidemiology, and End Results program of the National Cancer Institute. Source: Committee generated.

Mahboubi et al. (2013) published a large case–control study of lung cancer and formaldehyde exposure. The authors used a long-running study of lung cancer in Montreal that was based on incident cases gathered during two time periods: 1979–1986 and 1996–2002. The well-described exposure assessment methods were based on a detailed questionnaire on jobs and duties performed. Trained occupational hygienists evaluated each questionnaire, blinded to case and control status, on three dimensions of formaldehyde exposure: confidence (possible, probably, definite); relative concentration (low, medium, high); and frequency of use in a normal week (low, medium, high). The study was relatively large; there were 99 cases with exposure to formaldehyde that were judged by the occupational hygienists to be "substantial" exposures. The study found little to no evidence of incidence of lung cancer associated with any of the formaldehyde exposure measures. The study investigated potential confounding by smoking, and none was found. The study was able to evaluate effects separately in men and women, and no effect was observed in either gender. It was also able to stratify on the three primary histologic types of lung tumors (squamous cell, small cell, and adenocarcinoma) and, again, there was no evidence of an association with formaldehyde exposure for any type.

Siew et al. (2012) established a population-based cohort of all Finnish men who were born during 1906–1945 and followed the cohort for cancer incidence by linking to data in the Finnish Cancer Registry. They used the men's occupations reported to the 1970 national census to estimate occupational exposures to a wide array of chemicals, including formaldehyde, and found that men who developed lung cancer were 18% more likely to have jobs that involved exposure to formaldehyde than men who did not develop lung cancer (RR = 1.18, 95% CI 1.12–1.25). That finding was positive, and the size of the study (more than 30,000 lung-cancer cases) resulted in tight confidence limits, but the authors were doubtful of the finding because of the likelihood that they were unable to control fully for confounding by smoking and by concurrent exposures to other strong lung carcinogens, particularly asbestos. The committee concurred with those concerns.

Checkoway et al. (2011) had a strong study design, but the committee judged it to be weak for the purposes of this assessment because few cases were exposed to formaldehyde. The study was a large industrial case-cohort study (628 incidence lung-cancer cases) of Chinese female textile workers and it had detailed exposure assessment. However, the prevalence of formaldehyde exposure was low, and only two cases had 10 years or more of formaldehyde exposure. The resulting measure of association was imprecise: the hazard ratio for 10 or more years of formaldehyde exposure was 2.1 (95% CI 0.4–11).

Luo et al. (2011) conducted a population-based ecologic study of incident cases in US counties. They linked lung-cancer incidence from the Surveillance, Epidemiology, and End Results Program cancer registries to US Environmental Protection Agency Toxics Release Inventory data on formaldehyde emissions from industries. They found that a county's lung-cancer rate was positively associated with releases of formaldehyde (and chromium and nickel). For example, the RR was 1.18 (95% CI 1.05–1.33) when nonmetropolitan counties that had any formaldehyde release were compared with counties that had no formaldehyde release. The results are intriguing, but, as the authors note, evidence from individual-level studies is needed to support the finding.

The committee concluded that the newly identified studies do not provide enough evidence to indicate a causal association between formaldehyde and lung cancer. There remains a good possibility that confounding factors explain the increase in lung cancer reported in some formaldehyde studies. In addition, the studies yielded no epidemiologic evidence that indicated an association between formaldehyde exposure and cancer at other sites.

CANCER STUDIES IN EXPERIMENTAL ANIMALS

This section reviews the evidence of carcinogenicity in experimental animal studies and applies the NTP criteria to produce the committee's independent evaluation. In reviewing the evidence, the committee looked at primary literature and considered analyses in other reviews, including those by the International Agency for Research on Cancer (IARC 1982, 1995, 2006a) and NTP (2010, 2011). To capture studies that may have been published concurrently with the completion of the background document for formaldehyde up to 2013, the committee undertook an independent literature search. See Appendix D (Box D-2 and Figure D-2) for more information.

Studies of Low Power for Detecting Malignancies

Some bioassays discussed in the section "Studies of Cancer in Experimental Animals" of NTP's background document for formaldehyde are of limited adequacy to evaluate the carcinogenicity of formaldehyde (Table 3-8). Some of the studies were designed to follow up on studies that found carcinogenicity, for example, to explore hypotheses related to etiology or to look for differences in activity in different species. Those studies have findings of interest in considering progression to carcinogenesis, but they had low power to detect malignancy, mostly because they were not of sufficient duration. In addition, some studies have small groups, particularly the studies that used monkeys (Rusch et al. 1983; Monticello et al. 1989).

All the studies that were of low power to detect malignancies were inhalation studies except that by Tobe et al. (1989), which exposed animals to formaldehyde via drinking water. Tobe et al. had a relatively small group (20 male and 20 female) at the start of the study; all the animals in the high-dose group receiving 5,000 ppm of formaldehyde in drinking water and a substantial fraction in the low-dose groups receiving 200 ppm of formaldehyde in drinking water (46.9% of males and 33.7% of females) died before the end of the study, although survival in the group receiving 1,000 ppm of formaldehyde in drinking water was relatively good. Mortality began within the first month of the study. With the small initial group and substantial noncancer mortality in the high- and low-dose groups, the study has little overall power for evaluating the oral carcinogenicity of formaldehyde. Additional studies published decades ago that were identified from bioassay tabulations (for example, the US Public Health Service 149 series *Survey of Compounds Which Have Been Tested for Carcinogenicity*) were also of short duration and had other deficiencies (Garschin and Schabad 1936; Watanabe et al. 1954; Muller et al. 1978), as discussed in more detail in Chapter 2.

TABLE 3-8 Studies[a] of Low Power for Detecting Malignancies

Species	Limitations	Findings of Interest in Formaldehyde-Treated Animals	Reference
C3H mice	• Examined only lung; no examination of nose • Study terminated for most groups at 35 weeks • Small group in single animal group allowed to live longer	Basal-cell hyperplasia, epithelial stratification, squamous-cell metaplasia, and atypical metaplasia in trachea and major bronchi	Horton et al. 1963
Wistar rats	• Short duration (13 weeks) • Small group (10 male and 10 female)	Proliferative lesions in nasal and olfactory epithelium	Woutersen et al. 1987
Wistar rats	• Short duration (13 weeks) • Histopathology only of nasal cavity	Disarrangement, hyperplasia, squamous metaplasia with keratinization of epithelium	Wilmer et al. 1989
Wistar rats	• Short duration (1 year) • Small group (10 male) • Only nasal cavity examined	Increased basal-cell hyperplasia and squamous-cell metaplasia	Appelman et al. 1988
Wistar rats	• Relatively small initial group (20 male and 20 female) and high mortality	Forestomach hyperkeratosis, basal and squamous-cell hyperplasia; glandular stomach hyperplasia	Tobe et al. 1989
Wistar rats	• Short duration (32 weeks) • Small group (10 male)	8 of 10 treated rats with forestomach papilloma, none in controls	Takahashi et al. 1986
Fischer rats	• Short duration (26 weeks) • Relatively small group (20 male and 20 female)	Increased squamous-cell metaplasia and hyperplasia, basal-cell hyperplasia at high doses	Rusch et al. 1983
Syrian golden hamsters	• Short duration (26 weeks) • Small group (10 male and 10 female)	No significant findings	Rusch et al. 1983
Cynomolgus monkeys	• Short duration (26 weeks) • Small group (6 male) • Age unknown	Squamous-cell metaplasia and hyperplasia of nasal turbinates	Rusch et al. 1983
Rhesus monkeys	• Short duration (1–6 weeks) • Small group (9 male)	Mild degeneration and squamous-cell metaplasia of nasal epithelium; increased cell proliferation rate	Monticello et al. 1989

[a]All studies conducted by inhalation except studies by Tobe et al. (1989) and Takahashi et al. (1986), which were via drinking water.
Source: Committee generated.

The study by Takahashi et al. (1986), which exposed male Wistar rats to formaldehyde in water at 5,000 ppm for 32 weeks is notable. Although it was of short duration, eight of 10 exposed rats and no control animals developed forestomach papilloma. The formaldehyde group was serving as a reference group in a study of the effect of formaldehyde on N-methyl-N'-nitro-N-nitrosoguanidine carcinogenicity. Because of the very short study duration, the finding of tumors is particularly notable.

The two studies conducted in nonhuman primates are also noteworthy. They were of short duration and used small numbers of animals, but both studies demonstrated clear cellular and proliferative lesions of the nasal turbinates. Rusch et al. (1983) reported squamous-cell metaplasia and hyperplasia in the high-dose exposure group of six cynomolgus monkeys exposed to formaldehyde at 2.95 ppm 22 hours/day, 7 days/week for 26 weeks. Monticello et al. (1989) exposed rhesus monkeys to formaldehyde at 6 ppm 6 hours/day, 5 days/week for 1 week (n=3) or 6 weeks (n=3). The authors reported increased rates of nasal epithelial cell-proliferation with squamous-cell metaplasia of the transitional and respiratory epithelia of the nasal passages and squamous-cell metaplasia of the respiratory epithelia of the trachea and large airways of the bronchial tree. Even though those findings do not reflect overt carcinogenesis, they are highly reminiscent of the preneoplastic epithelial lesions of the nasal cavity that were observed to precede nasal malignancies in chronic rat studies.

Evidence from Informative Studies

Chapter 2 discusses whether the committee found NTP's evaluation of the evidence and application of its criteria scientifically sound. The committee's independent application of the NTP criteria emphasizes studies that are designed with greater sensitivity to detect an effect. Table 3-9 shows the highest-quality inhalation studies in boldface. They all had relatively large groups (90 animals or more), handled test material adequately, and included well-defined comparison groups (Kerns et al. 1983; Sellakumar et al. 1985; Monticello et al. 1996). The studies were all conducted in rats. In each, formaldehyde caused high incidences of rare malignant nasal tumors (squamous-cell carcinomas) at air-chamber concentrations of 10–15 ppm; these tumors are rarely seen in carcinogenesis bioassays and can be characterized as occurring "to an unusual degree" with respect to incidence. It is noteworthy that none of the animals in control groups in any of the long-term exposure studies had a tumor of this type. The Kerns et al. (1983) study was among the group of highest-quality studies. That experiment had a robust finding of squamous-cell carcinoma in both male and female rats, and the incidences were also increased to an unusual degree. The initial report of this study (Battelle 1981) stated there was a significant increase in bone marrow hyperplasia in rats following exposure to formaldehyde. The short-term exposure study by Feron et al. (1988) did not achieve statistical significance ($p = 0.1$ by Fisher exact comparison between the top dose group and controls).

TABLE 3-9 Nasal Squamous-Cell Carcinoma in Long-Term Inhalation Studies of Formaldehyde[1]

Species and Strain	Study Duration (week)[2]	Sex	Concentrations in Air (Incidences) No SCC Effect	SCC	Other Findings	Reference
Mouse B6C3F₁	104	M	0 ppm (0/109) 2 ppm (0/100) 5.6 ppm (0/106)	14.3 ppm (2/104)	Epithelial dysplasia and squamous metaplasia in high- and middle-dose groups; epithelial hyperplasia at high doses	Kerns et al. 1983; Battelle 1981
		F	0 ppm (0/114) 2 ppm (0/114) 5.6 ppm (0/112) 14.3 ppm (0/119)	—	Dysplasia in high- and middle-dose groups; squamous metaplasia in the high-dose group	
Rat Wistar	130 (13 weeks of exposure)[3]	M	0 ppm (0/45)	10 ppm (1/44) 20 ppm (3/44)	One carcinoma in situ and two polypoid adenomas at 20 ppm	Feron et al. 1988
	120	M	0 ppm (0/26)	0.1 ppm (1/26) 1 ppm (1/28) 10 ppm (1/26)	—	Woutersen et al. 1989
Rat F344	104	**M**	0 ppm (0/118) 2 ppm (0/118)	5.6 ppm (1/119) 14.3 ppm (51/117*)	Four high-dose animals with other nasal malignancies	Kerns et al. 1983
		F	0 ppm (0/114) 2 ppm (0/118)	5.6 ppm (1/116) 14.3 ppm (52/115*)	One high-dose female with other nasal malignancy	
	104	M	0 ppm (0/90) 0.7 ppm (0/90) 2 ppm (0/96)	6 ppm (1/90) 10 ppm (20/90*) 15 ppm (69/147*)	Nasal malignancies in one animal at 10 ppm and one animal at 15 ppm; polypoid adenomas in 14 animals at 15 ppm	Monticello et al. 1996
	120	M	0 ppm (0/32) 0.3 ppm (0/32) 2 ppm (0/32)	15 ppm (13/32*)	An additional 3 rats at 15 ppm with squamous-cell papilloma	Kamata et al. 1997

(Continued)

TABLE 3-9 Continued

Species and Strain	Study Duration (week)[2]	Sex	Concentrations in Air (Incidences)			Other Findings	Reference
			No SCC Effect	SCC			
Rats Sprague Dawley	Life	M	0 ppm (0/99)	15 ppm (38/100*)	**Two treated rats with other nasal malignancies; 10 with squamous-cell papillomas**	Sellakumar et al. 1985
	104	F	0 ppm (0/15)	12.4 ppm (1/16)	Squamous-cell metaplasia or dysplasia in 10 exposed rats	Holmström et al. 1989
Hamster Syrian Golden	Life	M	0 ppm (0/132) 10 ppm (0/88) 30 ppm (0/50)	—	Minimal hyperplastic and metaplastic response	Dalbey 1982

*Statistically significant, $p < 0.0001$ by pairwise Fisher exact comparison.

[1] Well-conducted studies with relatively large groups are in boldface.

[2] All exposures were for 6 hours/day, 5 days/week except the Dalbey (1982) study in hamsters, which had one group at 5 hours/day, 5 days/week and one group at 5 hours/day, 1 day/week.

[3] 13 weeks of exposure followed by a long period of no exposure. Results of experiments with shorter exposure times not tabulated.

Abbreviation: ppm, parts per million; SCC, squamous-cell carcinoma. Source: committee generated.

In addition to the findings of the robust rat studies, Kerns et al. (1983) carried out a study in mice. Nearly all 17 high-dose mice that survived 24 months had nasal lesions (dysplasia and metaplasia), and two also had squamous-cell carcinoma. As noted by the authors and in the background document for formaldehyde, that finding is sufficient to demonstrate the potential for these tumors in the mice exposed by inhalation when put into the context of evidence for this site in the rat and when the rarity of the tumor is considered. The findings of squamous-cell carcinoma in long-term studies that exposed mice and rats via inhalation are supported by the preneoplastic lesions (for example, squamous metaplasia with keratinization of epithelium) and other nasal lesions found in the shorter-term studies. The study using hamsters found no effect (Dalbey 1982).

The Kerns et al. (1983), Kamata et al. (1997), and Sellakumar et al. (1985) inhalation studies included histopathologic examinations of non–respiratory tract tissues; the other inhalation studies did not. Kerns et al. (1983) was reported in full in the Battelle (1981) report to the Chemical Industry Institute of Toxicology. The Battelle report discusses findings of leukemia and lymphoma that were not found to be exposure-related. However, diffuse multifocal bone marrow hyperplasia in formaldehyde-exposed animals was increased in both treated males (six of 114 controls vs 26 of 111 treated, $p = 0.0001$) and females (seven of 113 controls vs 28 of 115 treated, $p = 0.0001$). Kamata et al. (1997) and Sellakumar et al. (1985) reported no statistically significant nonrespiratory tumor findings but provided no detail regarding other non–respiratory tract histopathology.

The database for evaluating oral exposure to formaldehyde is less robust than for inhalation exposure. Three studies exposed rats to formaldehyde via drinking water over long periods (Til et al. 1989; Soffritti et al. 1989, 2002). The studies are described at length by IARC (2006a) and NTP (2010).

The study by Til et al. (1989) exposed Wistar rats to formaldehyde that was generated with 95% pure paraformaldehyde and 5% water. The administered drinking-water concentrations were 0, 20, 260, and 1,900 mg/L; the initial groups were 70 animals per sex at each dose; and the interim sacrifices occurred at 53 and 79 weeks. The intestines were not examined histologically in the middle- and low-dose groups but were in the high-dose group. The authors found no increases in cancer incidence in the gastrointestinal tract. A male in the low-dose group and a female in the control group had gastric papilloma. Nearly all male (seven out of 10) and female (five out of nine) animals in the highest-dose group had epithelial hyperplasia of the forestomach, and substantial fractions had focal hyperkeratosis of the forestomach and hyperplasia of the glandular stomach. In contrast, in the 32-week study by Takahashi et al. (1986), noted above in the discussion of the low-power studies, eight of 10 male Wistar rats exposed via drinking water to formaldehyde at 5,000 mg/L had stomach papilloma. The exposure level in the Takahashi et al. (1986) study was higher than in the Til et al. (1989) study.

In a series of experiments in Sprague Dawley rats, Soffritti et al. (1989, 2002) administered formaldehyde via drinking water. The studies included full histologic examination of all tissues. In the first study (Soffritti et al. 1989), formaldehyde of unspecified purity was administered to 25-week-old breeders (20 controls and 18 treated) at 2,500 mg/L in water. The offspring were exposed in utero via the dam and then postnatally via water for 104 weeks. In the breeders, no stomach or intestinal tumors were observed in the controls, whereas stomach tumors were observed in one treated female (benign) and one treated male (malignant). In the offspring, similarly, there were no stomach or intestinal tumors in the control animals (59 males and 49 females). However, in treated offspring (36 males and 37 females), a variety of benign and malignant gastrointestinal tumors were observed at a low incidence, including malignant leiomyosarcoma, which is exceedingly rare in these animals. Leiomyosarcoma was observed in stomach tissues in one treated female and one treated male and in intestinal tissue of five treated females (statistically significant at $p = 0.01$) (IARC 2006a, NTP 2010). In addition, nonleiomyosarcoma gastrointestinal tumors were observed in two males (one benign and one malignant) and one female (malignant).

Soffritti et al. (2002) later followed up with a long-term drinking-water study with multiple exposure groups and groups with lower exposures than in the earlier (Soffritti et al. 1989) study: 0, 10, 50, 100, 500, 1,000, and 1,500 mg/L; 50 animals of each sex per group, except for the controls, which had a group size of 100. Four treated males developed leiomyosarcoma at 10 mg/L (forestomach, one animal), 1,000 mg/L (glandular stomach, one animal), and 1,500 mg/L (intestine, two animals), and seven treated females developed leiomyoma at 10 mg/L (two animals), 50 mg/L (one animal), and 1,500 mg/L (three animals) or leiomyosarcoma at 50 mg/L (one animal). None of the 200 untreated control animals (100 male and 100 female) had these tumors.

Soffritti et al. (2002) also reported an increased incidence of hemolymphoreticular tumors in some groups. The finding is of interest, but there is uncertainty about it because of the changing counts of the tumors in earlier study reports (as noted by IARC 2006a), the pooling of tumors of different cellular origins, and recent questions raised about the evaluation of this class of tumors by this laboratory (Malarkey and Bucher 2011; Gift et al. 2013). Total mammary tumors also increased with increasing dose in the females; this, too, involved pooling of tumors of different origins (for example, adenocarcinoma and liposarcoma). Although noteworthy, the findings of hemolymphoreticular and mammary tumors are not used in the committee's independent evaluation.

Committee Evaluation in the Context of the Report on Carcinogens Listing Criteria

Applying the NTP criteria to the bioassay data for formaldehyde, the committee draws the following conclusions about exposure to formaldehyde in experimental animals:

1. Multiple species and multiple tissue types affected by the exposure:
   - Multiple species: Increase in malignant tumors in rats (F344 rats [Kerns et al. 1983; Monticello et al. 1996; Kamata et al. 1997], Sprague Dawley rats [Sellakumar et al. 1985; Soffritti et al. 1989], and Wistar rats [Feron et al. 1988; Woutersen et al. 1989]) and mice (B6C3F$_1$ mice [Kerns et al. 1983]).
   - Multiple tissue types: Malignancies of nasal epithelium (mostly squamous-cell carcinoma) (Kerns et al. 1983; Sellakumar et al. 1985; Feron et al. 1988; Woutersen et al. 1989; Monticello et al. 1996; Kamata et al. 1997) and gastrointestinal tract (leiomyosarcoma) (Soffritti et al. 1989 [offspring]; Soffritti et al. 2002 [adults]).
2. Carcinogenicity by multiple routes of exposure: Inhalation (Kerns et al. 1983; Sellakumar et al. 1985; Feron et al. 1988; Woutersen et al. 1989; Monticello et al. 1996; Kamata et al. 1997) and oral (Soffritti et al. 1989 [offspring]; Soffritti et al. 2002 [adults]).
3. Carcinogenicity to an unusual degree with respect to incidence, site, type of tumor, or age at onset: Nasal tumors are rare in untreated rats and in multiple studies occurred in treated rats at relatively high incidence (Kerns et al. 1983; Monticello et al. 1996).

The committee concludes that there is sufficient evidence that formaldehyde is carcinogenic in experimental animals.

TOXICOKINETICS

This section outlines multiple aspects of the toxicokinetics of gas-phase formaldehyde. The most likely route of exposure in humans is inhalation, and the committee has focused on this route. Information on the reactivity and metabolism of formaldehyde is followed by specific information on endogenous vs exogenous formaldehyde levels and on the inhalation dosimetry of this gas, particularly as related to the potential for absorption into the bloodstream and systemic distribution. The current report focuses on formaldehyde gas; however, it is worth noting that paraformaldehyde powder is used in some embalming and chemical applications. These uses may produce exposures to airborne particles of paraformaldehyde in addition to gas-phase formaldehyde. There is currently a dearth of information on human health effects associated with exposure to paraformaldehyde particles.

Reactivity and Metabolism

Formaldehyde is a volatile, organic, one-carbon aldehyde that exists as a gas at room temperature. It is water-soluble and reacts reversibly with water to form methanediol, which is the principle aqueous form in tissues after exposure to formaldehyde (Fox 1985). It can self-polymerize to form paraformaldehyde, which is a solid at room temperature that has the ability to break down when heated to release the monomer. It also reacts reversibly with amine and sulfhydryl groups, and this may ultimately result in cross-links between macromolecules. The inherent chemical reactivity of gas-phase formaldehyde is important to note because it plays a key role in its interaction with many macromolecules and cellular processes. The innate chemical reactivity of formaldehyde allows it to act as a cross-linking agent to fix tissue for pathological analysis and as a reactant in the synthesis of numerous industrial products. Those same chemical properties can, in part, explain its numerous toxic properties. Formaldehyde is reactive because its carbonyl atom acts as an electrophile, which reacts reversibly with nucleophilic sites on cell membranes, amino groups on proteins and DNA, and thiol groups on such biochemicals as glutathione (Bolt 1987).

The native reactivity of formaldehyde contributes to the well-established irritant properties of formaldehyde. Studies have found formaldehyde to cause dermatitis on dermal exposure and both eye and nasal irritation on inhalation exposure (Paustenbach et al. 1997). The nasal sensitization does not appear to be related to concentrations of glutathione–formaldehyde dehydrogenase; this indicates that formaldehyde itself, not metabolic products, is the irritant (Zeller et al. 2011b). Formaldehyde also reacts with macromolecules—a feature that has been used extensively to detect exogenous exposure to formaldehyde through measurement of formaldehyde–DNA adducts (ATSDR 1999; Lu et al. 2011) and proteins (Edrissi et al. 2013a). The reaction of formaldehyde with cellular components contributes to the sensitization of people to formaldehyde, which is manifested as allergic reactions and alterations in a person's immune system (Costa et al. 2013; Hosgood et al. 2013; Lino-dos-Santos-Franco et al. 2013). Although the mechanism is unclear, several reports associate formaldehyde with induction of an occupational asthmatic response in exposed people (Tang et al. 2009; McGwin et al. 2011) and in animal models (Wu et al. 2013).

Formaldehyde is rapidly absorbed and biotransformed extensively at the point of contact after ingestion or inhalation. It is primarily oxidatively biotransformed by glutathione-dependent formaldehyde dehydrogenase (FDH), officially named alcohol dehydrogenase 5 (ADH5), and S-formyl-glutathione dehydrogenase to formic acid (IARC 2006a). Formic acid can be ionized to formate and excreted via the kidney, further biotransformed to CO_2 and exhaled, or condensed with tetrahydrofolate and enter the one-carbon pool (IARC 2006a). In one study, 70% of a ^{14}C-labeled formaldehyde dose was found to be excreted as [^{14}C]CO_2 within 12 hours, and the remainder entered the one-carbon pool, where it was incorporated into biomolecules in the body (Buss et al. 1964). Formaldehyde dehydrogenases are ubiquitous in all tissues, including the respiratory tract, with no distinct "regional" differences in the biotransformation of formaldehyde (Casanova-Schmitz et al. 1984; Thompson et al. 2008). The biotransformation of formaldehyde is similar in all species tested. The rapid biotransformation of formaldehyde at the point of contact limits the access of formaldehyde systemically.

Endogenous vs Exogenous Sources

Formaldehyde exposure has both exogenous and endogenous sources. It is produced intracellularly as a component of the one-carbon pool intermediary metabolism pathways. It is also the product of metabolism of drugs and other exogenous compounds (NTP 2010; NRC 2011). Because formaldehyde is normally present in tissues, the toxicokinetics of exogenous formaldehyde exposure must be evaluated in the context of the relatively large amounts of formaldehyde (near 0.1 mM) that are endogenously present. Measurement of tissue formaldehyde is somewhat difficult because of its volatility and reactivity. Many techniques rely on extraction followed by mass spectrometry (for example, Heck et al. 1982). Those methods provide a measure of free and reversibly bound formaldehyde but do not differentiate between the two. Formaldehyde, through the one-carbon pool, is metabolically incorporated into tissue macromolecules. Therefore, simple use of ^{14}C-labeled formaldehyde does not provide a direct measure of the distribution of parent exogenously administered formaldehyde (NTP 2010; NRC 2011). As noted above, because of its reactivity, formaldehyde may form DNA–protein cross-links, DNA–DNA cross-links, and protein or DNA adducts (Lu et al. 2010a; NTP 2010; NRC 2011; Edrissi et al. 2013b). Those moieties have the advantage of being more stable and longer-lasting than formaldehyde itself and have been used as biomarkers of cellular exposure to formaldehyde. It is important to recognize that use of the moieties (for example, DNA–protein cross-links) as biomarkers of cellular formaldehyde delivery does not require a direct link to tumorigenesis.

The endogenous formaldehyde concentration in whole blood of rodents and nonhuman primates is about 0.1 mM. The concentration in tissues is probably somewhat higher (NTP 2010; NRC 2011). That value represents free plus reversibly bound formaldehyde. Information on the fraction of blood formaldehyde that is free vs bound is not available. Whether from endogenous or exogenous sources, formaldehyde is extensively metabolized to formate via formaldehyde dehydrogenase as described above.

Inhalation Dosimetry

Because inhalation is the most likely route of exposure to formaldehyde, an understanding of the fate of inhaled formaldehyde is critical for evaluation of its toxicity. As would be expected for a water-soluble highly reactive gas (Kimbell 2006), inhaled formaldehyde is effectively removed from the airstream. Thus, it is expected that formaldehyde will be efficiently removed from the airstream in the first airways with which it comes into contact, either the nose during nose breathing or the tracheobronchial airways during mouth breathing. Water-soluble reactive gases may be absorbed efficiently in the mouth and pharynx during mouth breathing (Frank et al. 1969); although this is likely to occur with formaldehyde, it has not been confirmed experimentally. Experimental studies in the dog (Egle 1972) indicate greater than 95% deposition of inhaled formaldehyde in the nose, lower respiratory tract, and total respiratory tract. A published abstract (Patterson et al. 1986) provides similar data on nasal deposition in the rat.

Numerous state-of-the-art inhalation dosimetry mathematical models have been directed toward dosimetry of inhaled formaldehyde. They have recently been extensively and appropriately reviewed (NRC 2011). The models suggest that inhaled formaldehyde is not deposited uniformly throughout the nose, but local areas, "hot spots", receive a higher delivery of the dose than other areas. Those areas correlate closely, in the rat, with areas in which DNA–protein cross-link studies indicate high cellular delivery and with areas in which tumors are most likely to arise. Models suggest that rates of localized delivery to small regions in the human nose may be similar to those observed in rats exposed at the same concentration (Kimbell et al. 2001). The modeling prediction adds weight to the idea that formaldehyde may pose a carcinogenic hazard to the human nose. Models suggest that, despite the existence of localized hot spots within the nose, nasal deposition efficiency averaged over the entire nose is lower in humans or nonhuman primates than in rats, leading to greater penetration of inhaled formaldehyde to the lower respiratory tract. That is supported by DNA–protein cross-links studies that suggest higher cellular delivery of inhaled formaldehyde to the trachea and mainstream bronchi in nonhuman primates than in rats (Heck et al. 1989; Casanova et al. 1991). Unlike the obligate nose-breathing rodent, humans are capable of mouth breathing; this would greatly increase the delivery of inhaled formaldehyde to the lower airways.

The airway epithelium is metabolically active. Of relevance to formaldehyde disposition within nasal tissues is the presence of ADH5/FDH. The metabolic pathways offer an effective clearance mechanism for formaldehyde. Only formaldehyde that escapes metabolism is available for binding to tissue macromolecules or potentially available for absorption into the blood. Like all metabolic pathways, formaldehyde metabolism demonstrates saturation kinetics. As saturation occurs, the likelihood of reaction of formaldehyde with tissue macromolecules or of penetration of formaldehyde to deeper tissues increases. On the basis of modeling efforts and DNA–protein cross-link assessments, saturation kinetics may occur at concentrations above 2 ppm in the rodent nose. Specifically, a nonlinear relationship between inspired concentration and DNA–protein cross-links in the nose is observed at exposure concentrations of 6 ppm or higher, greatly exceeding what would be expected for a linear increase from the DNA–protein cross-links observed at concentrations of 2 ppm or lower (NTP 2010; NRC 2011).

Absorption into Blood

The disposition of formaldehyde in airway tissues and distribution throughout the body are important for understanding the potential for tissue injury in airways or distant tissues. As previously noted, formaldehyde reacts readily and reversibly with sulfhydryl and amine moieties. Formaldehyde reacts reversibly with water to form methanediol, with the equilibrium strongly favoring methanediol. As outlined by Georgieva et al. (2003), it is not likely that the dissociation of methanediol to form formaldehyde is rate-limiting (in contrast with the reaction with macromolecules), so this process is not critical for determining formaldehyde disposition in nasal tissues (NRC 2011). Because formaldehyde reactions are reversible, it is possible that an individual formaldehyde molecule, if it is not metabolically degraded, may shuttle from one binding site to another. Therefore, an individual *endogenous* formaldehyde molecule could be distributed away from its site of formation, and an individual *exogenous* formaldehyde molecule could be distributed to tissues away from its site of first contact. That would occur only if the formaldehyde molecule escaped metabolic transformation. Because ADH5/FDH is ubiquitously expressed, including expression in red blood cells, the likelihood of metabolic transformation is high, and this lowers the likelihood of penetration to distant tissues through the bloodstream.

Anatomic features of the airways are highly relevant to the potential for absorption into the blood and systemic distribution of formaldehyde (NRC 2011). The air–blood barrier of the nose and large tracheobronchial airways consists of a mucous lining layer overlying a pseudostratified columnar mucociliary epithelium. Residing below the basement membrane, the submucosal space of the nasal airways is highly vascularized. In the nose, a superficial capillary layer is present just below the basement membrane (Figure 3-3). This relationship is important for evaluation of formaldehyde disposition in the nose. Presumably, the target cells for tumorigenesis in the nasal airways are the basal cells that reside on the basement membrane. Immediately below the basement membrane are the vessels of the superficial capillary layer of the nose. The total epithelial thickness in the nose depends on the site but is generally less than 0.05 mm in rodents and humans (Schroeter et al. 2008). A similar structure exists with respect to the nasal associated lymphoid tissue (NALT), which resides just below the basement membrane (Figure 3-3).

On the basis of mathematical modeling and estimation of the rates of reaction and metabolism, it has been estimated that formaldehyde would penetrate to some depth in nasal tissues (see Figure 3-4) (Georgieva et al. 2003). Specifically, the modeling efforts suggest that the formaldehyde concentration at the depth of 0.05 mm (below the basement membrane) is greater than 50% of the concentration at the mucus–tissue interface. Thus, the concentration–tissue depth profile appears to have a shallow slope. Formaldehyde is clearly cytotoxic to the nasal epithelium, and the nasal epithelial basal cells are probably the target for nasal tumorigenesis; this indicates that reactive formaldehyde penetrates to this depth in the nose. Given the shallow slope of the concentration–tissue depth profile, it is likely that toxicologically significant concentrations of formaldehyde penetrate somewhat deeper to the superficial capillary layer of the nose, inasmuch as these capillaries are adjacent to the basement membrane and basal

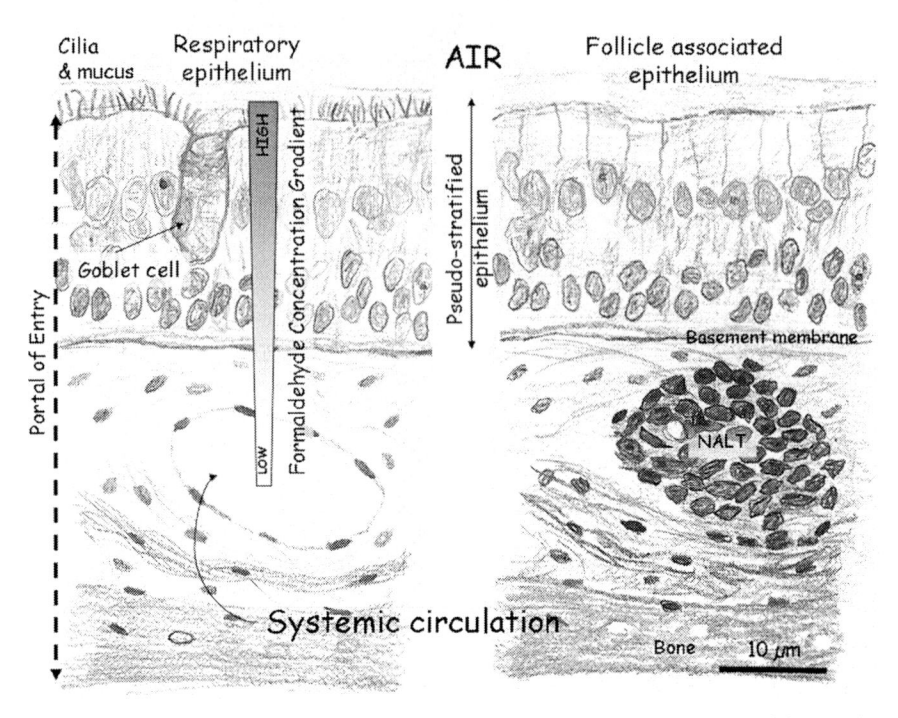

FIGURE 3-3 Schematic representation of the structure of the nasal mucosa of the respiratory epithelium and follicle-associated epithelium. For both epithelia, a concentration gradient for exogenous formaldehyde during inhalation exposure will exist with concentrations at the superficial layer (closest to the airstream) being higher than concentration in deeper layers. As outlined in the text, this gradient is due to the reaction of formaldehyde with tissue substrates or metabolism via ADH5/FDH. It is worth noting that basal cells, a target for formaldehyde-induced carcinogenesis, lie immediately above the basement membrane and capillaries and nasal associated lymphoid tissue (NALT) lie immediately below the basement membrane. Source: NRC 2011, p. 32.

cells (see above). Thus, at sufficient airborne concentrations, biologically significant concentrations of formaldehyde may be present in the nasal submucosa and capillary bed. It should be recognized, however, that the presence of formaldehyde in the nasal submucosa and capillary bed does not itself indicate that biologically significant concentrations of formaldehyde penetrate via the bloodstream to distant tissues. A toxicokinetic approach could be formulated to estimate the exposure concentrations that would be required to raise systemic blood formaldehyde substantially above endogenous concentrations. To the committee's knowledge, that has not been performed.

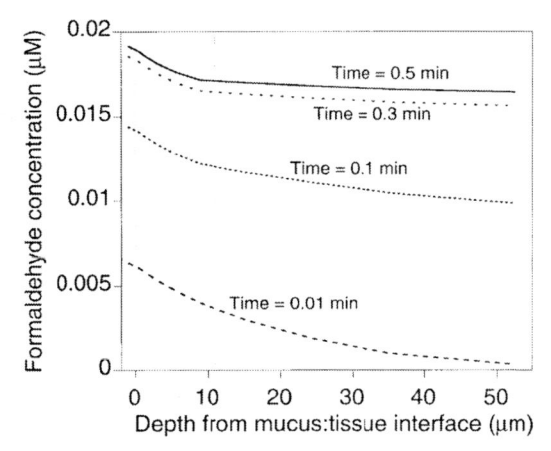

FIGURE 3-4 Model-based estimates of exogenous formaldehyde concentration in nasal tissues during inhalation exposure to 6 ppm formaldehyde. Tissue concentrations increase quickly from 0.1 to 0.5 minutes after the onset of exposure as a quasi-steady state is established. Readily apparent is the prediction that the formaldehyde concentration at a depth of 50 μm, measured from the mucus:tissue interface, is fairly similar to the concentration at the interface itself. Source: Georgieva et al. 2003. Reprinted with permission; copyright 2003, *Inhalation Toxicology*.

Distribution of Inhaled Formaldehyde

The nose receives about 1% of cardiac output, and mathematical models suggest that about one-third of nasal circulation (0.33% of total cardiac output) may perfuse the superficial capillary layer (Gloede et al. 2011). Venous blood from the nose is ultimately mixed with the systemic venous blood. On the basis of relative perfusion rates, blood from the entire nose is diluted by a factor of 100 (because the nose receives 1% of the cardiac output) with systemic venous blood; blood from the superficial capillary layer is diluted by a factor of about 300 with systemic venous blood before distribution to the body. From that perspective, it can be appreciated that although the concentration of an inhaled xenobiotic in the nasal capillary blood may be high, its concentration is greatly reduced (by a factor of 100–300) as blood from the nose mixes with systemic venous blood. The underlying structure of the large tracheobronchial airways is similar to that of the nose; thus, the relationships described above are qualitatively similar for the lower airways. The entire tracheobronchial tree receives about 1% of cardiac output (Gloede et al. 2011). As for the nose, any xenobiotic absorbed into the tracheobronchial circulation of the large airways is diluted by a factor of about 100 as the venous output from the airways mixes with the systemic venous blood.

Although it is theoretically possible that an individual exogenous formaldehyde molecule could be distributed away from the portal of entry, mass-balance and kinetic arguments and experimental data strongly suggest that this does not occur to a great extent. Specifically, multiple studies that used different conceptual approaches, from simple mass-balance estimates (Heck and Casanova 2004; Nielsen et al. 2013) to more detailed pharmacokinetic analysis (Franks 2005), universally support the conclusion that the amount of formaldehyde that is inhaled (at reasonable exposure concentrations) and absorbed into circulation is much lower than the endogenous amounts in circulation. Analytic studies did not observe a large increase in the total content of formaldehyde in blood or tissue above the endogenous concentrations during inhalation exposure (NTP 2010; NRC 2011). Published literature, relying on gas chromatographic and mass spectrometry techniques, indicates that blood formaldehyde (measured as free plus reversibly bound) is not increased in the rat, monkey, or human by inhalation exposure to formaldehyde (Heck et al. 1985; Casanova et al. 1988). Studies that use bound formaldehyde as a biomarker and that rely on dual-labeled formaldehyde also did not observe an increase in tissue formaldehyde during inhalation exposure in any tissue except the nose (Lu et al. 2011; Moeller et al. 2011; Edrissi et al. 2013b). Contrary to these findings are findings of formaldehyde adducts in the blood of exposed individuals. One study reported increases in blood albumin–formaldehyde adducts in workers exposed to formaldehyde (Pala et al. 2008); another reported increases in formaldehyde–hemoglobin adducts (Bono et al. 2006). Mass-balance arguments call the validity of those findings into question (Nielsen et al. 2013), specifically that the amount of formaldehyde that would be required to raise albumin adducts or hemoglobin adducts to the levels reported is much greater than the amount that was inhaled.

Recent well-designed studies have relied on dual labeled formaldehyde to measure formaldehyde–DNA adducts as a biomarker of delivered dose of *exogenous* formaldehyde for comparison with *endogenous* concentrations (Lu et al. 2010a,b; Moeller et al. 2011). They indicate that endogenous formaldehyde–DNA adducts are ubiquitous throughout the body. Increased exogenous formaldehyde–DNA adducts are observed in nasal tissues of rodents and nonhuman primates after inhalation exposure to formaldehyde, and this validates the sensitivity of the technique. High concentrations of exogenous formaldehyde–DNA adducts are not observed in distal tissues, including bone marrow, after formaldehyde inhalation. Those experiments provide strong evidence that formaldehyde exposure at the concentrations used (up to 15 ppm) does not result in substantial delivery of exogenous formaldehyde to nonrespiratory tissues. The results have recently been confirmed by using formaldehyde–lysine adducts as biomarkers instead of formaldehyde–DNA adducts (Edrissi et al. 2013b).

MECHANISMS OF CARCINOGENESIS

The mechanisms of carcinogenesis of formaldehyde have been the subject of intense research for decades, and a large evidence base is available from which to draw inferences and conclusions. Despite the wealth of information available on a variety of test systems, from naked DNA (that is, DNA without any associated proteins) to experimental animals and exposed humans, it is still being debated what mechanistic events take place in tissues that have been suggested as targets for formaldehyde-associated carcinogenesis. Such debate is informed, in large part, by the considerations of formaldehyde toxicokinetics, inasmuch as formaldehyde is both a highly reactive molecule and an endogenously formed compound produced in the course of normal cellular metabolism. There is evidence that exogenously administered formaldehyde is responsible for noncancer and cancer effects at the portal of entry, such as nasal mucosa or other parts of the upper aerodigestive tract, depending on the mode of administration and breathing patterns. It has been more controversial whether formaldehyde itself or products of its biotransformation may reach tissues that do not come into direct contact with inhaled or ingested formaldehyde in experimental animals or humans, and a detailed discussion of the available evidence is provided under the section "Toxicokinetics" above. There is general agreement that systemic delivery of formaldehyde is unlikely (NRC 2011), but it is also true that various toxicity phenotypes (for example, genotoxicity and mutagenicity in circulating blood cells, changes in the number of circulating cells and bone marrow cells, and gene expression changes in blood) have been found in cells and tissues that are not in direct contact with exogenously administered formaldehyde. That apparent inconsistency notwithstanding, the committee concurs with the conclusions drawn by the National Research Council Committee to Review EPA's Draft IRIS Assessment of Formaldehyde (NRC 2011) that it is important to differentiate between systemic delivery of formaldehyde and systemic effects. It is possible that the "systemic delivery of formaldehyde is not a prerequisite for some of the reported systemic effects seen after formaldehyde exposure. Those effects may result from indirect modes of action associated with local effects, especially irritation, inflammation, and stress" (NRC 2011, p. 36).

The present committee found that the most sensible characterization of the adverse health effects of formaldehyde and associated mechanisms is that proposed by NRC (2011). Specifically, a wide array of the adverse outcomes that have been associated with formaldehyde exposure are best classified into portal-of-entry and systemic categories, which are defined as follows: *portal-of-entry* effects are effects that arise from direct interaction of inhaled or ingested formaldehyde with the affected cells or tissues; *systemic* effects are effects that occur beyond tissues or cells at the portal of entry. The committee notes, however, that it is plausible that some of the systemic effects, most notably genotoxicity in circulating blood cells, may have resulted from the exposure of these cells at the portal-of-entry tissues (for example, lymphoid tissue in the nasal mucosa).

As discussed in previous sections, the committee relied on the background document for formaldehyde, published reviews, and assessments performed by other authoritative bodies to ensure that relevant literature was captured up to the publication of the 12th RoC. It also considered literature, comments, and arguments provided during its open session and submitted by other sources during the duration of the study. The committee carried out its own literature search (see Appendix D) for publications that are pertinent to the major postulated modes of carcinogenic action of formaldehyde (genotoxicity, cell proliferation and apoptosis, and effects on the immune system). The committee's exclusion criteria and detailed search strategies for studies related to genotoxicity and mutagenicity are presented in Box D-3 and for studies related to immune effects are presented in Box D-4. Literature trees were used to document identification and selection of the literature evidence (Figures D-3 and D-4). The general question that the committee addressed was, What is the evidence that the following mechanistic events—genotoxicity and mutagenicity or effects on the hematologic system—are part of the overall mode of action of formaldehyde-associated carcinogenicity? The outcomes of the searches and the evidence available in the background document for formaldehyde (NTP 2010) were evaluated together and are detailed below.

The committee notes that because of the limitations of time and resources several of the mechanisms that have been proposed by NTP (2011) to explain the carcinogenicity of formaldehyde (such as cytotoxicity followed by compensatory proliferation and oxidative stress) have not been evaluated by conducting new literature searches. In the course of the review of the substance profile for formaldehyde in the NTP 12th RoC (see Chapter 2), the committee found that the mechanism of cytotoxicity followed by compensatory cell proliferation is a well-established portal-of-entry mechanism that is not controversial. On the contrary, oxidative stress is a mechanistic event that has not been addressed in detail and on which the evidence base is too small to draw firm conclusions. The committee focused its attention on the mechanistic evidence that is related to genotoxicity and mutagenicity, hematologic effects, and data from toxicogenomic studies, which reflects broad biologic responses and is thus informative as both the overall effect and specific pathways that may be perturbed by exposure to formaldehyde.

The RoC does not present quantitative assessments of risks of cancer associated with the substances listed. Therefore, the committee did not explicitly take into consideration the issue of the dose or concentration of formaldehyde that was applied or evaluated in each study. The background document for formaldehyde contains extensive information on the doses and concentrations used in various studies, and, where it is available, the committee notes dose-dependent and time-dependent trends in the new studies that have been published since June 10, 2011.

Finally, the committee notes that although the mode of action of a chemical substance is an important component of decision-making to protect human health, the guidelines established by various national and international agencies that conduct such assessments differ in how such information is gathered, presented, and evaluated (Box 3-1). The guidance documents of IARC, the US Environmental Protection Agency (EPA), and the International Programme on Chemical Safety (IPCS) are informative, but the committee's charge (see Appendix B) was to integrate the level-of-evidence conclusions and to consider all relevant information in *accordance with RoC listing criteria*. In that respect, for each listed substance, the RoC includes studies of genotoxicity and of biologic mechanisms. The listing criteria are used to guide the evaluation of the human, animal, and mechanistic evidence. The listing criteria specifically state that "data derived from the study of tissues or cells from humans exposed to the substance in question, which can be useful for evaluating whether a relevant cancer mechanism is operating in humans" (NTP 2010, p. iv), constitute one of the lines of evidence used to support whether there is *sufficient* or *limited* evidence of carcinogenicity from studies in humans.

BOX 3-1 Guidance from Various Agencies on the
Use of Mechanistic and Other Relevant Data

The IARC Monographs Program operates under the general guidance of a preamble, which specifies that a working group is to consider mechanistic and other relevant data because they "may provide evidence of carcinogenicity and also help in assessing the relevance and importance of findings of cancer in animals and in humans" (IARC 2006b, p. 15). The preamble outlines "scientific principles, rather than a specification of working procedures" (p. 1), for the experts who participate in the development of each monograph. It notes that "the procedures through which a Working Group implements these principles are not specified in detail" (p. 1).

The EPA *Guidelines for Carcinogen Risk Assessment* (EPA 2005) state that the agency's assessments should discuss the available information on the modes of action and associated key events of chemicals under evaluation. Specifically, the assessments aim to address several questions pertaining to the extent and quality of the evidence on the hypothesized mode of action. The questions include sufficiency of supporting information from test animals, relevance to humans, and any information that may suggest that particular populations or life stages can be especially susceptible to the hypothesized mode of action. It is noted, however, that "in the absence of sufficiently, scientifically justifiable mode of action information, EPA generally takes public health-protective, default positions regarding the interpretation of toxicologic and epidemiologic data" (EPA 2005, p. 1-10).

IPCS developed a mode-of-action relevance framework for the analysis of mechanistic evidence on chemical carcinogens in experimental animals and its relevance to humans (Boobis et al. 2008). The framework calls for determining whether the weight of evidence based on experimental observations is sufficient to establish a hypothesized mode of action. A series of key events causally related to the toxic effect are then identified using an approach based on the Bradford Hill criteria and compared qualitatively and quantitatively between experimental animals and humans.

Genotoxicity and Mutagenicity

The data available to examine the potential role of genotoxicity and mutagenicity of formaldehyde are extensive. Those effects are likely to be relevant for all cancer sites that have been associated with formaldehyde exposure. Nearly all aspects of genotoxicity and mutagenicity have been studied with formaldehyde, so assertive conclusions can be drawn from the available evidence.

The committee collated the evidence on all the mechanistic events that make up the genotoxic mode of action into separate tables (see Appendix E). In each table, the committee separated studies by type of the model system, including a clear division between the portal-of-entry and systemic effects in in vivo studies. Publications that have evaluated a particular mechanistic event and found evidence supporting or refuting each were included. In addition, a summary table (Table 3-10) was constructed to present the totality of the evidence available on each mechanistic event in each experimental model system.

Overall, the evidence on genotoxicity and mutagenicity of formaldehyde resulted from studies that evaluated DNA adducts (Table E-1), DNA–DNA cross-links (Table E-2) and DNA–protein cross-links (Table E-3), DNA strand breaks (Table E-4), mutations (Table E-5), sister-chromatid exchanges (Table E-6), micronuclei (Table E-7), and chromosomal aberrations (Table E-8). Several published studies have also examined the DNA-repair responses to formaldehyde-induced DNA damage. Owing to the paucity of data, the model systems used in these studies, and the scope of the present committee's charge, that information was not included in the evaluation. Similarly, the committee found that although some reports examined the possible role of genetic polymorphisms in the genotoxic potential of formaldehyde or ensuing adverse outcomes, the overall database was not robust and did not provide strong evidence that human variability factors (genetic polymorphisms) may be critical for drawing conclusions. All studies included in Appendix E were examined in full text (including translations, where applicable) by at least two committee members, who independently determined whether a given study observed an important effect or lack thereof with respect to the phenotype named in each table. Studies were categorized as positive if a statistically significant effect was observed. Studies were categorized as negative if the results reported an absence of a particular effect (that is, no statistically significant difference from the appropriate control group). Although the committee members exercised their scientific judgment in categorizing studies and determining their relevance to each phenotype, the committee did not perform a formal quality assessment of each individual study, whether it was categorized as positive or negative. The committee members also did not make judgments about the study design or methodology, recognizing that all the studies had been subjected to some form of peer review before publication.

TABLE 3-10 Summary of Published Studies on the Genotoxic and Mutagenic Effects of Formaldehyde in Test Systems and Organisms[1]

		DNA Adducts	DDX	DPX	Strand breaks	Mutations	SCE	MN	CA
Cellfree systems		+ (7/0)	+ (3/0)	+ (3/0)					
Nonmammalian model organisms					+ (6/0)	+**			
Mammalian in vitro	Rodent	+ (1/0)		+ (14/1)	+ (6/2)	+/- (3/2)	+ (9/0)	+ (4/0)	+ (5/0)
	Human	+ (2/0)		+ (23/0)	+ (8/0)	+ (6/0)	+ (6/0)	+ (4/0)	+ (6/2)
Mammalian in vivo: portal-of-entry effects	Rodent	+ (2/0)		+ (8/0)	- (0/1)	+/- (1/1)		-/+ (1/2)	+ (1/0)
	Primate	+ (1/0)		+ (2/0)					
	Human							+/- (11/3)	
Mammalian in vivo: systemic* effects	Rodent	- (0/1)		+/- (2/2)	+/- (2/1)	+ (1/0)	- (0/2)	-/+ (4/5)	- (2/5)
	Primate	- (0/1)		- (0/2)					
	Human	+# (1/0)		+ (3/0)	+ (9/2)		-/+ (7/9)	+ (18/3)	+/- (11/5)

[1]Total numbers of studies demonstrating effect or lack thereof are indicated in parentheses. See Appendix E for data that support this summary table: DNA adducts (Table E-1), DNA–DNA cross-links (Table E-2), DNA–protein cross-links (Table E-3), DNA strand breaks (Table E-4), mutations (Table E-5), sister-chromatid exchanges (Table E-6), micronuclei (Table E-7), and chromosomal aberrations (Table E-8).
+: all or most of the studies indicate the effect.
+/-: most of the studies indicate the effect, although many show lack thereof.
-/+: most of the studies indicate lack of the effect, although many positive studies have been published.
-: all or most of the studies indicate lack of the effect.
*The committee acknowledges that although most investigators consider the effects on circulating-blood mononucleated cells as systemic because cells for the analyses were collected from the systemic circulation, it is plausible that the cells had been exposed to formaldehyde in the nose through lymphoid tissue in the mucosa.
**The results are overwhelmingly positive for point mutations and overwhelmingly negative for frame-shift mutations.
#M1G adduct has been postulated to be the result of secondary DNA damage caused by formaldehyde-associated oxidative stress.
Abbreviations: DNA, deoxyribonucleic acid; DDX, DNA–DNA cross-links; DPX, DNA–protein cross-links; SCE, sister-chromatid exchanges; MN, muconuclei; CA, chromosomal aberrations. Source: Committee generated.

The committee's work was informed by the Bradford Hill criteria (Hill 1965) for determining causality between exposure to formaldehyde and findings of genotoxicity and mutagenicity. Although those criteria have been proposed for determinations of causality in epidemiologic studies, they do not all apply to the evaluation of the mechanistic evidence. As noted in EPA guidelines (EPA 2005, p. 2-13), "one . . . cannot simply count up the numbers of studies reporting statistically significant results or statistically non-significant results for carcinogenesis and related MOAs [modes of action] and reach credible conclusions about the relative strength of the evidence and the likelihood of causality." Thus, the committee, upon systematizing the available mechanistic evidence pertaining to the genotoxicity and mutagenicity of formaldehyde into tables, appraised the evidence by using the general guidance of the "causal criteria" (EPA 2005) to determine its overall strength for drawing conclusions about causality for each of the mechanistic events identified in the tables. Because the body of evidence on genotoxicity and mutagenicity of formaldehyde is very large, the mechanistic synthesis does not contain many citations to the individual publications; all the evidence is presented in multiple tables.

Owing to the challenge of establishing whether and how formaldehyde can exert point-of-entry and systemic effects, the committee chose to evaluate causality for each of the mechanistic events in three broad categories:

1) Effects on the naked DNA or on the DNA of nonmammalian organisms or mammalian cells in vitro.

2) Effects observed on the portal-of-entry tissues of animals or humans exposed to formaldehyde.

3) Systemic effects in animals or humans exposed to formaldehyde.

The latter two are most relevant to the determination of the cancer-hazard classification according to the RoC listing criteria, which call for conclusions to be based on the information "derived from the study of tissues or cells from humans exposed to the substance in question" (NTP 2011, p. 198). Again, the committee acknowledges that although most investigators consider the effects on circulating blood mononucleated cells to be systemic because cells for the analyses were collected from the systemic circulation, it is plausible that these cells have been exposed to formaldehyde in the nose through lymphoid tissue in the mucosa.

Effects of Formaldehyde on Naked DNA or on DNA of Nonmammalian Organisms or Mammalian Cells in Vitro

The totality of the evidence overwhelmingly shows that when formaldehyde is added to naked DNA or nonmammalian organisms or mammalian cells are incubated in the presence of formaldehyde, DNA adducts (Table E-1), cross-links (Tables E-2, E-3), strand breaks (Table E-4), mutations (Table E-5), and clastogenic damage (Tables E-6, E-7, and E-8) are found. Studies were conducted in different types of model systems and have produced consistent results.

The evidence of genotoxicity and mutagenicity of formaldehyde comes from studies where different model systems were tested and various molecular techniques were used to evaluate the effects. Because all studies evaluated in this category used formaldehyde, specificity of the effects being caused by formaldehyde has been firmly established. In addition, many studies used appropriate positive and negative controls, and this further strengthens the specificity of the association. The temporal relationship of the observed association is clear in that the studies evaluated genotoxic and mutagenic effects after DNA or cells came into contact with formaldehyde. Dose–response relationships between genotoxic and mutagenic effects and formaldehyde were observed in studies that had appropriate designs. For example, DNA–protein cross-links were formed in a concentration–response manner in human lymphoblastoid cell lines (Ren et al. 2013), epithelium-like human lung cells (Speit et al. 2010), and isolated human lymphocytes (Neuss et al. 2010a,b). Similar observations were made in whole-blood cultures for sister-chromatid exchanges, micronuclei, and chromosomal aberrations (Schmid and Speit 2007; Ren et al. 2013).

The committee concludes that the genotoxic and mutagenic mode of action of formaldehyde in studies of naked DNA, studies of DNA from nonmammalian organisms, and studies of mammalian cells in vitro is consistent, strong, and specific to the formaldehyde exposure. Both temporal and dose–response relationships have been established. This mechanistic event is relevant to human cells because all the genotoxic effects observed in studies of naked DNA, nonmammalian model organisms, or cells from rodents have been also observed in human cells, either established cell lines or primary cells.

Effects on the Portal-of-Entry Tissues of Animals or Humans Exposed to Formaldehyde

Because various studies reviewed by the committee may have used different routes of administration of formaldehyde and because of the differences in breathing patterns among rodents and humans, the committee considered the following anatomic regions as points of entry: nasal passages, oral cavity and upper aerodigestive tract, and forestomach (in gavage studies). The committee identified no studies that evaluated DNA–DNA cross-links or sister-chromatid exchanges in exposed rodents or humans at the portal of entry, so these mechanistic events were not considered in this section.

Most of the evidence of genotoxic and mutagenic effects at the portal of entry, depending on the end point studied, is from studies of laboratory rodents and exposed humans. Several reports evaluated pertinent mechanistic events in nonhuman primates. Studies of DNA adducts (Table E-1), even though the database is not large, showed that formaldehyde-induced DNA damage is consistently observed in both rodents (Lu et al. 2010a, 2011) and nonhuman primates

(Moeller et al. 2011). Similarly, consistent evidence from a large number of studies of rodents and nonhuman primates demonstrates formation of DNA–protein cross-links (Table E-3). Positive and negative findings, albeit from a small number of studies of formaldehyde exposure of rodents, are equally divided for strand breaks (Table E-4), mutations (Table E-5), micronuclei (Table E-7), and chromosomal aberrations (Table E-8). In humans exposed to formaldehyde, formation of micronuclei was examined in cells at the portal of entry, and 11 of 14 studies demonstrated a positive association (Table E-7). Overall, the findings are consistent with genotoxic and mutagenic effects of formaldehyde observed in naked DNA, in the DNA of nonmammalian organisms, and in mammalian cells in vitro.

Evidence of genotoxicity and mutagenicity of formaldehyde in exposed humans is strong, even though several studies reported no induction of micronuclei. The positive observations were made in studies of diverse groups of subjects that were exposed to formaldehyde. Various assays have been used to evaluate the mechanistic events, and statistical significance of the effects was established in the positive studies.

In rodent and nonhuman primate studies, formaldehyde exposures were well documented (for example, purified reagent-grade formaldehyde was used). Furthermore, several studies of DNA damage have used ^{13}C-labeled formaldehyde (Lu et al. 2010a, 2011; Moeller et al. 2011), which shows that the genotoxic effects of formaldehyde occur at the portal of entry. In human studies, many investigators established the association between formaldehyde and these mechanistic events through exposure monitoring, albeit most of the studies were of occupational cohorts and the presence of other agents cannot be excluded. Some of the studies that found no evidence of micronuclei in portal-of-entry tissues from humans (Speit et al. 2007; Zeller et al. 2011a) is evidence that questions the association in controlled exposures of volunteers to formaldehyde.

Studies of rodents and nonhuman primates provide strong evidence for a temporal relationship of the observed association because the genotoxic and mutagenic effects were observed after exposure to formaldehyde. In many human studies, temporality was established by collecting samples before and after exposure in the workplace.

Studies of rodents and nonhuman primates provide strong evidence of concentration–response relationships in the genotoxicity of formaldehyde at the portal of entry (Lu et al. 2010a, 2011; Moeller et al. 2011). The concentrations of formaldehyde used in the studies (around 1–10 ppm) are comparable with or an order of magnitude higher than those documented in human occupational exposures. The shape of the concentration–response curve of several biomarkers of genotoxicity in the portal-of-entry tissues in rodents is nearly identical with that for tumorigenesis in the noses of rodents (Swenberg et al. 2013).

The committee concludes that the genotoxic and mutagenic mode of action of formaldehyde in the portal-of-entry tissues of animals or humans exposed to formaldehyde is supported by the experimental evidence. Several negative studies notwithstanding, the evidence is consistent, strong, and specific with respect to an association following exposure to formaldehyde. Both temporal and exposure–response relationships have been established, most strongly in the studies of experimental animals (rodents and nonhuman primates). This mode of action is relevant to humans because statistically significant increases in the number or frequency of micronuclei, known biomarkers of clastogenesis, have been observed in most, but not all, of the studies of portal-of-entry tissues from humans exposed to formaldehyde.

Systemic Effects in Animals or Humans Exposed to Formaldehyde

Systemic effects are effects that occur outside cells or tissues that come into direct contact with exogenous formaldehyde. Most studies in the systemic-effects category examined genotoxic and mutagenic effects of formaldehyde in circulating blood mononucleated cells unless stated otherwise. The committee acknowledges, however, that although most investigators consider the effects on circulating blood mononucleated cells as systemic because cells for the analyses were collected from the systemic circulation, it is also plausible that these cells were exposed to formaldehyde in the nose through lymphoid tissue in the mucosa.

Most of the experimental evidence that is available for drawing conclusions about systemic genotoxic and mutagenic effects of formaldehyde comes from studies in humans exposed to formaldehyde, mostly in occupational settings. Fewer experimental-animal (for example, rodent) studies have been conducted, and only two studies of nonhuman primates examined some of the mechanistic events in question. Overall, the database pertaining to this question is most consistent in exposed humans in whom formaldehyde exposure-associated DNA–protein cross-links (Table E-3), strand breaks (Table E-4), micronuclei (Table E-7), and chromosomal aberrations (Table E-8) were detected in most of the studies. Data on sister-chromatid exchange formation in response to exposure to formaldehyde in humans are almost equally divided for and against (Table E-6). In studies in rodents, there is little positive evidence of clastogenic effects of formaldehyde on circulating blood cells but some evidence of strand breaks and mutations. Studies of nonhuman primates found no evidence of the increased formation of DNA adducts in bone marrow after exogenous administration of ^{13}C-labeled formaldehyde (Moeller et al. 2011) or the presence of DNA–protein cross-links in the most distal regions (lung parenchyma) of the respiratory tract (Casanova et al. 1991).

Evidence of genotoxicity and mutagenicity of formaldehyde in exposed humans is strong because various assays were used to evaluate these effects, data come from a number of independent laboratories around the world, and the positive studies were conducted on humans exposed in a variety of occupational settings (for example, pathologists, embalmers, and anatomy students). The negative human studies also contribute important information in that the diversity of the study designs and occupational and laboratory-based exposures is appreciable.

The studies of rodents and nonhuman primates used controlled exposures to purified reagent-grade formaldehyde, and some studies even used controlled exposures to ^{13}C-labeled formaldehyde, which increases the specificity of the negative observations. Human studies were largely in occupational exposure scenarios in which formaldehyde was the primary—not the only—agent and other chemical (for example, solvent) or physical (for example, wood-dust) exposures were possible. Formaldehyde-associated DNA–protein cross-links were found in three human studies (Table E-3); however, most of the end points that were evaluated in the positive studies, such as strand breaks (Table E-4) and clastogenic effects (Tables E-6, E-7, and E-8), are difficult to attribute specifically to formaldehyde. Thus, the specificity of the observed positive associations is somewhat uncertain.

In many—not all—positive human studies, a temporal relationship was established by collecting samples before and after exposure in the workplace (Lin et al. 2013) or by considering the extent of employment in an occupation in which formaldehyde exposure is very likely (Viegas et al. 2010; Ladeira et al. 2011; Souza and Devi 2014). Some studies of rodents and nonhuman primates provide strong evidence of lack of a dose–response relationship in the formation of exogenous formaldehyde-induced DNA adducts (Lu et al. 2010a, 2011; Moeller et al. 2011). Recent studies that evaluated DNA–protein cross-links, however, show dose-dependent increases in this biomarker of genotoxicity in tissues (bone marrow, liver, spleen, and testes) that are not in direct contact with inhaled formaldehyde (Ye et al. 2013). Some of the positive human studies found a relationship between the clastogenic effects of formaldehyde and exposure duration (Viegas et al. 2010; Ladeira et al. 2011; Souza and Devi 2014) or dose (Jiang et al. 2010).

The committee concludes that the systemic genotoxic and mutagenic mode of action of formaldehyde is sufficiently supported by the evidence from studies of humans exposed to formaldehyde. The committee acknowledges that reporting bias against negative results could be a limitation of its approach to reviewing the mechanistic evidence (NRC 2014); however, that limitation does not detract from the conclusion that formaldehyde can induce systemic genotoxic changes. The evidence is consistent and strong, albeit it is difficult to establish unequivocal specificity of the effects following exposure to formaldehyde in the human studies. Whereas the committee recognizes some inconsistencies among data in experimental animals and humans and among genotoxicity biomarkers, this variability does not undermine the committee's conclusion. Both temporal and exposure–response relationships have been demonstrated in studies of humans exposed to formaldehyde. This mode of action is relevant to humans because most of the positive evidence comes from studies of humans exposed to formaldehyde. The data do not exclude the possibility of other modes of action but strongly suggest a causal relationship between exposure to formaldehyde and human cancer.

Hematologic Effects

The systemic effects of formaldehyde exposure and the association with hematopoietic malignancies have been a source of debate, and there has been much interest in the hematologic effects of formaldehyde exposure. Several recent studies have evaluated the effects of formaldehyde on circulating hematopoietic cells, and a number of them were published after the release of the NTP 12th RoC. In this section, the committee focuses on changes in hematopoietic-cell number or function—that is, "hematologic effects". It did not consider genotoxicity studies and studies of altered gene expression because they are covered in other sections of this chapter. In addition, given that few studies have been designed to address the clinical significance of hematologic effects, to address the mechanisms by which hematologic effects may arise after exposure, or to address mechanisms that contribute to adverse health effects (including cancer), these topics were not considered by the committee. The focus of this section is on evaluation of recently available evidence related to the hematologic effects of formaldehyde in human and animal exposure studies and evidence that is available from in vitro studies.

Hematologic Effects in Humans Exposed to Formaldehyde

Hematologic effects of formaldehyde include effects on cells of the hematopoietic system that are circulating in the peripheral blood, are present in hematologic tissues (such as bone marrow, lymph nodes, and spleen), or are present in other tissues, whether at the portal of entry or not. The available data primarily reflect the hematologic consequences of exposure to inhaled formaldehyde in humans without addressing the mechanism or health consequences of the findings.

Many studies have addressed the hematologic effects of exposure to formaldehyde in humans (Tables 3-11 and 3-12). Six studies that examined inhalation exposures of formaldehyde in humans reported decreases in overall white blood cells, and three reported decreases in red cells and platelets. Studies have also reported many other hematologic effects, such as increases in monocytes, eosinophils, and some T-cell subsets and decreases in neutrophils and T-cell function. It should be noted that several studies have reported contrasting findings in the same hematologic characteristic, such as increases vs decreases in total lymphocyte concentration and T-, B-, and NK-cell subsets. Given that formaldehyde exposure concentrations, durations, and sources varied greatly among studies, it is difficult to reconcile those results. However, taken as a whole, the body of evidence demonstrates consistently that exposure of humans to inhaled formaldehyde is associated with an array of hematologic effects.

TABLE 3-11 Recent Studies of Hematologic Effects of Formaldehyde[a]

Model	Subjects	Exposure	Sample	Main Hematologic Findings (Excluding Genotoxicity)[b]	Reference
Inhalation exposure in humans	Workers (43 formaldehyde-exposed, 51 age- and sex-matched controls)	Factory workers exposed to formaldehyde–melamine resins compared with workers without formaldehyde exposure; mean formaldehyde exposure 1.28 (0.63–2.51) ppm vs <0.03 ppm	Peripheral blood tested for lymphocyte subsets	Extension of Zhang et al. (2010) using the same subjects and reporting additional assays. Total NK-cell and T-cell counts were 24% and 16% lower, respectively, in exposed workers. Decreased counts in exposed workers were observed for CD8+ T cells, CD8+ effector memory T cells, and regulatory T cells. B-cell numbers did not differ significantly.	Hosgood et al. 2013
	Workers (43 formaldehyde-exposed, 51 age- and sex-matched controls)	Factory workers exposed to formaldehyde–melamine resins; exposures same as Hosgood et al. (2013)	Peripheral blood measures (complete blood count and WBC differential)	Reanalysis of Zhang et al. (2010) data. Differences in blood measures when examined in context of population averages for Chinese and general populations and when controlled for potential confounders (for example, suspected thalassemia trait) suggest that effects attributed to formaldehyde are not clinically significant. Concerns were raised regarding relevance of CFU-GM assays to AML stem-cell biology.	Gentry et al. 2013
	Male workers (46 formaldehyde-exposed, 46 controls)	Factory workers in two medium-density fiberboard-producing plants; measured formaldehyde levels; 8-hour TWA = 0.20 ± 0.06 ppm (0.10–0.33 ppm)	Blood samples measured for lymphocyte subsets, immunoglobulins, complement proteins, and TNFα concentrations	Percentage of lymphocytes was increased 13% in formaldehyde-exposed workers. Absolute numbers and percentages of T cells (17% and 6%, respectively) and NK cells (48% and 34%, respectively) were higher, IgG (23%) and IgM (27%) in exposed workers were statistically lower, TNFα was significantly higher (308%). No significant differences in white blood cell, erythrocytes, hemoglobin, neutrophils, or monocytes were observed.	Aydin et al. 2013
	Workers (35 formaldehyde-exposed, 35 controls)	Pathology anatomy workers with >1 year exposure in four hospitals in Portugal	Blood sample measured for lymphocyte subsets (T, B, and NK cells) and MN, SCE, and TCR mutations	Overall, 30% decrease in percentage of B cells (CD19+) found in formaldehyde-exposed workers compared with controls ($p < 0.05$). Decreased B-cell percentage was significant in multivariate analysis	Costa et al. 2013

		and nonexposed administrative workers in same facilities; 8-hour TWA mean exposure = 0.36 ± 0.03 ppm (range 0.23–0.69 ppm)		(including sex, smoking, and age) (p = 0.014). T cells (CD3+) and helper T cells (CD3+/CD4+) increased when analyzed by formaldehyde exposure (p = 0.002 and 0.006, respectively) and in multivariate analysis (p = 0.024 and 0.037, respectively). NK cells (CD16+/CD56+) decreased on basis of individual exposure levels (p < 0.001) and in multivariate analysis (p < 0.001).	
	Female workers (37 formaldehyde-exposed, 37 controls)	Workers, formaldehyde-exposed women in four pathology departments in Hungary; 8-hour TWA mean exposure = 0.9 mg/m³ measured in three of four sites; 16 subjects identified as having exposure to organic solvents in addition to formaldehyde were analyzed separately	Blood samples measured for apoptosis, proliferation, HPRT function, UV-induced DNA synthesis, CA, SCE, and T-cell activation marker CD71 after PHA stimulation in vitro	Apoptotic cells after PHA stimulation were mean of 77% higher in formaldehyde-only exposed workers compared with controls. Lectin labeling index and variant frequency, measures of HPRT function, were significantly increased and decreased, respectively, in formaldehyde-exposed workers. CD71 expression on T cells and BrdU incorporation were not significantly changed.	Jakab et al. 2010
Inhalation exposure in animals	Male Balb/c mice	Inhaled formaldehyde at 0, 0.5, 3 mg/m³, 8 hours/day, 5 days/week (5 days on, 2 days off), 13 days	Blood measured for complete blood count (cell types and hemoglobin), BM for histology, ROS, GSH, cytochrome 1A1, GSTT1, NFκB, TNFα, and IL-1b	Formaldehyde exposure led to a significant decrease (p<0.05) in white blood cells, red blood cells, and lymphocytes after exposure to 0.5 mg/m³ of formaldehyde (43%, 7%, and 39%, respectively), and 3.0 mg/m³ of formaldehyde (52%, 27%, and 43% respectively). Platelet counts were significantly increased (p<0.05) after exposure to formaldehyde at 0.5 mg/m³ (109%) and 3.0 mg/m (67%). Monocytes and granulocytes were not significantly changed. At a formaldehyde exposure of 0.5 mg/m³ and 3.0 mg/m³, ROS levels in BM increased by 31% and 102%, respectively; CYP1A1 increased by 8% and 37%, respectively; and GSTT1 decreased by 0% and13%,	Zhang et al. 2013

(Continued)

149

TABLE 3-11 Continued

Model	Subjects	Exposure	Sample	Main Hematologic Findings (Excluding Genotoxicity)[b]	Reference
Inhalation exposure in animals				respectively. At 3.0 mg/m³ of formaldehyde, NFkB increased by 34%, and inflammatory cytokines were increased—TNFα by 42% and IL-1b by 98%.	
	Female C57BL/6	Inhaled formaldehyde at 0, 5, 10 ppm, 6 hours/day, 5 days/week, 14 days of exposure	BM, lymph node, spleen, liver, and lung measured for cell types and NK function	Formaldehyde-exposed mice showed 30% increase in percentage of T cells (CD3+), 38% increase in CD8+ T cells, and 28% decrease in B cells (B220+) in spleen at 10 ppm, but absolute numbers were not significantly different. No change in percentage of CD4+ or CD8+ T cells in BM, lymph nodes, liver, or lung. Percentage of NK cells (NK1.1+) in lung was decreased in concentration-dependent manner (decrease of 19% at 5 ppm and 58% at 10 ppm) and returned nearly to normal in 2 weeks after last formaldehyde exposure. Absolute numbers of NK cells were reduced in lung, but total leukocyte numbers were not changed at 10 ppm. Total number of cells present in BAL was increased >20-fold in formaldehyde-exposed mice, but absolute number of NK cells was decreased by over 65%, as were Ly49 receptor expression levels on NK cells. Similarly, percentage and total NK cells and Ly49 expression were decreased in spleen in a time-dependent manner, but no change in total splenocytes was observed. IFNg, perforin, and CD122 were decreased in NK cells from lung and spleen of formaldehyde-exposed mice, and LPS-mediated increase in these proteins was inhibited after formaldehyde exposure in lung. NK cytolytic activity (chromium release assay) of splenic NK cells was decreased at 2–3 weeks of formaldehyde exposure. Decrease in NK-cell numbers (approximately 30%) and function were seen in tumor-bearing mice exposed to formaldehyde. Decreases in NK viability and differentiation in vitro were also observed.	Kim et al. 2013

	Outbred female white rats	Inhaled formaldehyde at 12.8 ± 0.69 mg/m³, 4 hours/day, 5 days/week, 10 weeks	Blood measured for blood cell types, hemoglobin, MN, and multiple serum proteins and amino acids	Of blood-cell types and hemoglobin, formaldehyde-exposed rats had statistically significant differences ($p < 0.05$) in percentage of lymphocytes (11% increase) and percentage of segmented neutrophils (31% decrease).	Katsnelson et al. 2013
	Female Wistar rats	Inhaled formaldehyde, nebulized at 0.32%, 90 minutes/day for 3 consecutive days	Blood and bone marrow samples measured for cell subsets; BAL fluid leukocytes	Sham-control rats were part of a larger study of female sex hormone effects on formaldehyde-induced airway inflammation. Formaldehyde exposure in these control rats showed a 111% increase in WBC, including mononuclear and neutrophil subsets in BAL fluid. Sham-control rats had 197% increase in WBC, but there was >70% decrease in BM cell numbers in formaldehyde-exposed rats. >19-fold increase in degranulated mast cells was seen in lungs of formaldehyde-exposed control rats.	Lino-dos-Santos-Franco et al. 2011
In vitro studies	Primary expanded human erythroid progenitor cells from PBMCs	0–150 mcM formaldehyde in tissue culture	Cell growth and cell cycle distribution	Formaldehyde exposure suppressed in vitro human erythroid progenitor cell expansion in dose-dependent manner.	Ji et al. 2013
	Primary expanded human NK cells from PBMCs	0–3,200 µM formaldehyde in tissue culture examined at 10, 30, 60, and 120 minutes	Morphology, viability, apoptosis, cytotoxicity (killing tumor-cell activity), cytokine and cytolytic proteins, and secretion of NK cells were evaluated	NK-cell viability, cytolytic activity, and perforin secretion were decreased above 800 micromolar.	Li et al. 2013
	Primary mouse BM MSCs	0–200 mcM formaldehyde in tissue culture	Viability (MTT assay)	BM MSCs demonstrated cytotoxicity >75 micromolar.	She et al. 2013
	Human lymphoblastoid cell lines	0–200 mcM formaldehyde for 24 hours in tissue culture	Viability (AnnexinV binding and PI staining)	FANCD2-deficient lymphoblastoid cell line was statistically more sensitive to formaldehyde-induced cell death than FANCD2-expressing control.	Ren et al. 2013

(Continued)

151

TABLE 3-11 Continued

Model	Subjects	Exposure	Sample	Main Hematologic Findings (Excluding Genotoxicity)[b]	Reference
	Primary human lymphocytes from 30 volunteers	0–1.152 mg/mL formaldehyde after PHA stimulation for 72 hours	Viability (trypan blue and MTT assay)	Statistically significant decreases in viability seen at formaldehyde concentrations above 0.036 mg/mL.	Pongsavee 2011

[a]The studies in this table were identified through the committee's literature search. See Appendix D for more details of the search.
[b]All reported findings are significant with $p < 0.05$.

Abbreviations: AML, acute myeloid leukemia; B, bursa-derived cells; BAL, bronchoalveolar lavage; BM, bone marrow; BrdU, bromodeoxyuridine; CA, chromosomal aberrations; CD, cluster of differentiation; CFU-GM, colony-forming unit-granulocyte-macrophage; CYP1A1, cytochrome P450, family 1, subfamily A, polypeptide 1; DNA, deoxyribonucleic acid; FANCD2, fanconi anemia group D2 protein; GSTT1, glutathione s-transferase theta 1; HPRT, hypoxanthine-guanine phosphoribosyltransferase; GSH, glutathione; IFNg, interferon gamma; IgG, immunoglobulin G; IgM, immunoglobulin M; IL-1b, interleukin-1 beta; LPS, lipopolysaccharide; Ly49 - killer cell lectin-like receptor subfamily A; mg/m^3, milligram per cubic meter; mg/mL, milligrams per milliliter; MN, micronucleus test; MSC, mesenchymal stem cell; MTT, methylthiazol tetrazolium; NFkB, nuclear factor kappa-light-chain-enhancer of activated B cells; NK, natural killer cells; PBMC, peripheral blood mononucleated cell; PHA, phytohemagglutinin; ppm, parts per million; ROS, reactive oxygen species; SCE, sister-chromatid exchange; T, thymus cells; TCR, T-cell receptors; TNFa, tumor necrosis factor alpha; TWA, time-weighted average; UV, ultraviolet; WBC, white blood cell count. Source: Committee generated.

TABLE 3-12 Studies Grouped by Hematologic Effects

Model	Cell Type	Hematologic Effects[a]	Reference
Inhalation exposure in humans	WBC	↓ Total WBC	Qian et al. 1988; Kuo et al.1997; Targ and Zhang 2003; Cheng et al. 2004; Tong et al. 2007; Zhang et al. 2010
		↑ Percentage of lymphocytes	Aydin et al. 2013
		↓ Total lymphocytes	Zhang et al. 2010
		↓ CFU formation	
	T cells	↓ Total T cells and CD8+ T cells	Ying et al. 1999; Ye et al. 2005; Hosgood et al. 2013
		↓ CD4+ T cells	Ying et al. 1999
		↑ CD4/CD8 ratio	Ying et al. 1999; Ye et al. 2005
		↓ CD26+ activated T cells	Madison et al. 1991
		↑ T cells	Aydin et al. 2013; Costa et al. 2013
		Impaired mitogen-induced proliferation of lymphocytes	Vargova et al. 1992
		↑ PHA-induced apoptosis	Jakab et al. 2010
	NK cells	↓ NK cells	Costa et al. 2013; Hosgood et al. 2013
		↑ NK cells	Aydin et al. 2013
	B cells	↓ B cells percentage	Ying et al. 1999; Ye et al. 2005
		↓ B cell percentage	Costa et al. 2013
		↑ autoantibodies and anti-FA-albumin conjugates	Madison et al. 1991
		↓ IgM/IgA	Qian et al. 1988
		↓ IgG/IgM	Aydin et al. 2013
	Erythrocytes	↓ erythrocyte count and hematocrit level	Lyapina et al. 2004
		↓ hemoglobin level	Yang 2007
		↑ MCV	Zhang et al. 2010
	Neutrophils	↓ spontaneous respiratory burst activity	Lyapina et al. 2004
		↑ susceptibility to infection	

(Continued)

153

TABLE 3-12 Continued

Model	Cell Type	Hematologic Effects[a]	Reference
	Monocytes	↑ monocytes in indoor FA+nitrogen dioxide exposure	Erdei et al. 2003
	Eosinophils	↑ eosinophils	Qian et al. 1988
	Platelets	↓ platelets	Tong et al. 2007; Yang 2007; Zhang et al. 2010
Inhalation exposure in animals	WBC	↓ WBC	Brondeau et al. 1990; Zhang et al. 2013
		↑ WBC	Lino-dos-Santos-Franco et al. 2011
		↓ lymphocytes	Zhang et al. 2013
		↓ lymphocyte viability	Pongsavee 2011
		↑percentage lymphocytes	Kim et al. 2013
		↓ bone marrow cell numbers	Lino-dos-Santos-Franco et al. 2011
		↑ bone marrow cell numbers	Battelle 1981
	T cells	↑ percentage of T cells and CD8+ T cells	Kim et al. 2013
	NK cells	↓ total and percentage of NK cells ↓ IFNg, perforin, and CD122 in NK cells. ↓ cytolytic activity and NK differentiation ex vivo	Kim et al. 2013
	B cells	↓ B cells	Kim et al. 2013
	Neutrophils	↓ segmented neutrophils	Katsnelson et al. 2013
	Erythrocytes	↓ erythrocytes	Zhang et al. 2013
	Platelets	↑platelets	Zhang et al. 2013
In vitro studies	T cells	↓ IFNg and IL-10 in stimulated human T cells	Sasaki et al. 2009
	B cells	↓ viability of human lymphoblastoid cells	Ren et al. 2013
	NK cells	↓ NK cell viability, cytolytic activity, and perforin secretion	Li et al. 2013
	Erythrocytes	↓ expansion of human erythroid progenitor cells in vitro	Ji et al. 2013
	MSCs	↓ viability of bone marrow stromal cells	She et al. 2013

[a]All significant effects reported with p <0.05.
Source: Committee generated.

Given the variability of blood measures in any person over time and the heterogeneity among people in a population, it is difficult to find statistically significant changes in blood measures in human studies. Thus, it is notable that despite the inherent limitations of studying hematologic measures, over 14 recently published studies reported statistically significant hematologic effects on multiple hematopoietic-cell types. Although there are valid concerns about some results in individual studies (for example, the authors of one study used the consequences of the thalassemia trait for mean corpuscular volume to explain the findings), it is unlikely that most of these studies have been confounded by such issues. In light of the numerous studies that have reported significant differences in multiple measures, there is a strong association between inhaled formaldehyde exposure in humans and hematologic effects.

Although confounding exposures may complicate the interpretation of some studies, most of the studies documented efforts to identify possible confounding factors. Several studies were conducted in occupations in which formaldehyde was probably the predominant exposure during the period of study. One study showed that hematologic changes occurred in individual subjects over a limited period of exposure (Ying et al. 1999). Thus, the hematologic effects observed in those studies establish a specific association with inhaled formaldehyde in humans. Establishing the temporal relationship of exposure and effect is difficult in most human-exposure studies. Several studies report an association between duration of employment and exposure to formaldehyde, and an 8-week anatomy-laboratory exposure study (Ying et al. 1999) supports a temporal relationship. There is evidence from one human study that supports a biologic gradient of formaldehyde exposure and hematologic effects. In this study, increases in T cells and decreases in NK cells were proportional to formaldehyde exposure level (Costa et al. 2013). Those findings are supported by findings in animal-exposure studies (see below).

Hematologic Effects in Animals Exposed to Formaldehyde

Experimental-animal studies are informative with regard to the specificity, temporal relationship, and exposure–response relationship between formaldehyde and hematologic effects. It can be argued that rodents and humans differ in the mechanics of inhalation, the physiology of hematopoietic-cell turnover, and DNA-repair mechanisms. Therefore, results of animal studies were evaluated as supporting data, whereas the human data presented above are considered the primary source of evidence of potential associations of formaldehyde exposure and hematologic effects.

Six studies addressed the hematologic effects of exposure to formaldehyde in animals in vivo, of which four were published after the publication of the 12th RoC (Tables 3-11 and 3-12). There is poor agreement between individual studies as to the direction of hematologic effects induced by inhaled formaldehyde in animals. In particular, increased or decreased effects on total white-cell counts, total lymphocyte counts, and percentage limit the ability to interpret the results. In addition, other hematologic effects have been reported in only one study, so the consistency of the findings cannot be assessed. The committee finds limited evidence of consistent hematologic effects in the few available studies of formaldehyde-exposed animal models alone.

In the experimental-animal studies, the associations that were observed were often strong in magnitude or level of statistical significance, although the clinical and biologic significance is unknown (Katsnelson et al. 2013; Kim et al. 2013). Thus, the strength of those specific associations is quite high, even if the consistency of the findings is limited. As is expected in experimental-animal studies, the observed multiple hematologic effects can be closely linked to the tested agent, and this establishes a specific association with formaldehyde. By their nature, the animal-exposure studies establish the temporal relationship between inhaled formaldehyde exposure and multiple hematologic effects. In particular, specific hematologic effects were shown to depend on the duration of exposure (Kim et al. 2013). Two animal studies reported multiple hematologic measures, and effects on them were proportional to formaldehyde concentrations (Kim et al. 2013; Zhang et al. 2013). The results suggest an exposure–response relationship between formaldehyde exposure and hematologic effects.

Hematologic Effects on Isolated Animal or Human Cells

In vitro studies of hematologic effects are of limited utility because they evaluate a nonintact hematopoietic system, which ignores the complex interplay between various cell types and the vascular and lymphohematopoietic organs. Such studies do not account for the complex dynamics between the portal of entry and the systemic distribution of formaldehyde.

The committee examined six studies that reported cytotoxic effects on or functional consequences for hematopoietic cells or bone marrow stromal cells, of which five were published after publication of the 12th RoC (Table 3-10 and 3-11). All six studies reported deleterious effects of formaldehyde exposure on T cells, B cells, NK cells, or bone marrow stromal cells; this suggests that formaldehyde may have hematologic effects if it comes into direct contact with these cell types. However, given the unclear relevance of direct exposure in in vitro studies, particularly exposure to formaldehyde, the committee concludes that although the available literature demonstrates a deleterious effect of formaldehyde exposure on hematologic cells in vitro, it is difficult to draw firm conclusions regarding the hematologic effects of formaldehyde on isolated animal and human cells. The direct effects reported on several hematopoietic cell types raise important questions, but additional studies are needed that account for the physiologic exposure of hematopoietic cells to formaldehyde and its metabolites and for poorly understood systemic consequences.

Conclusions and Considerations for Hematologic Effects

The committee concludes that the association of inhalation formaldehyde exposure and diverse hematologic effects is supported by evidence from human studies. Studies in experimental animals provide some additional support. The consistency of individual hematologic effects varied among multiple human and animal studies, and many reported decreases in hematologic measures. The strength of the association in multiple reports of hematologic effects in multiple populations is convincing. The specificity of findings in exposed humans is challenging, but select human studies and experimental-animal studies support the specificity of the association. The temporal relationship is adequately addressed in most studies, and the biologic gradient is addressed in some studies, particularly in animal studies. Taken as a whole, the body of evidence from studies of exposed humans and animals indicates broad and strong associations between exposure to inhaled formaldehyde and hematologic effects.

Toxicogenomics

Toxicogenomics is the study of gene-expression changes elicited by a toxicant. The committee reviewed recent toxicogenomic publications to gain a better understanding of changes in gene expression after formaldehyde exposure. The committee looked specifically at toxicogenomic studies and identified eight publications that had microarray data. Those publications provided information on the genomewide expression of mRNA transcripts in humans, experimental animals, or cultured cells after exposure to formaldehyde. Five of the publications were identified through the committee's independent literature search for genotoxicity and mutagenicity studies (Andersen et al. 2010; Zeller et al. 2011a; Cheah et al. 2013; Neuss et al. 2010b; Kuehner et al. 2013) (see Figure D-4), and two additional publications were identified from the reference lists of those relevant publications (Hester et al. 2003; Andersen et al. 2008). One publication was identified during the committee's secondary ad hoc effort to identify relevant literature (Rager et al. 2013). Five of the eight publications described exposures in humans or experimental animals (Hester et al. 2003; Andersen et al. 2008, 2010; Zeller et al. 2011a; Rager et al. 2013), and the remaining three used cell culture (Hester et al. 2003; Neuss et al. 2010b; Cheah et al. 2013). The eight studies are described in more detail in this section and in Table 3-13.

Zeller et al. (2011a) used volunteer human subjects to examine transcriptomal changes in nasal inferior turbinate biopsies and peripheral blood samples after inhalation of formaldehyde vapor at up to 0.8 ppm 4 hours/day for 5 days. This is the only study that the committee identified that attempted to examine both portal-of-entry and systemic transcriptomal effects of formaldehyde. The authors reported that 27 mRNA transcripts were differentially expressed between exposed and nonexposed conditions in the nasal specimens. In

TABLE 3-13 Transcriptomal Profiling Studies

Model	Subjects	Exposure	Sample	Criteriaa	Main conclusions	Reference
Animals or humans	Human volunteers: male nonsmokers or ex-smokers	Formaldehyde vapor Up to 0.8 ppm 4 hours/day for 5 days 3 groups (5–8/group)	Before and after exposure (paired) nasal biopsy (inferior turbinate); venous whole blood	2-fold or 1.5-fold; $p < 0.05$ (paired t); no FDR correction	Formaldehyde exposure affected mRNA expression in nasal biopsy or blood samples only marginally. There were 2–17 and 25–67 differentially expressed genes identified in biopsies with 2.0- and 1.5-fold difference criteria, respectively. Results identified 0–9 and 6–39 differentially expressed genes in the blood with 2.0- and 1.5-fold difference criteria, respectively. Differentially expressed genes identified in the three exposure groups showed little overlap. No significant specific pathways involving differentially expressed genes were apparent. When FDR cutoff (less than 10%) was applied in addition to 1.5-fold change cutoff, no differentially expressed genes were detected.	Zeller et al. 2011a
	Nonhuman primates: male Cynomolgus macaques	Formaldehyde vapor 0 (n = 2), 2 (n = 3), and 6 ppm (n = 3) 6 hours/day for 2 days	Nasal epithelial tissue from maxilloturbinate region collected by necropsy	1.5-fold; $p < 0.05$ (ANOVA); FDR corrected $q < 0.1$	Low (2 ppm) and high (6 ppm) doses of formaldehyde changed 3 and 13 micro-RNA expressions, respectively. Suppression of transcriptional targets of most significantly increased miRNA (miR-125b) was confirmed by real-time PCR. Induction of transcriptional targets of most robustly decreased miRNA (miR-142-3p) was also confirmed by real-time PCR. Four miR-125b targets encoding proapoptotic regulators BAK1, CASP2, MAP2K7, and MCL1 [b] were downregulated. Thus, formaldehyde	Rager et al. 2013

				exposure disrupts miRNA expression in nasal epithelium and probably affects apoptosis.	
Rats: male F344/CrlBR	Formaldehyde vapor 0, 0.7, 2, 6, 10, and 15 ppm 6 hours/day for 1, 4, 13 weeks (15 per dose per time)	Nasal surface epithelial cells (lateral meatus and nasoturbinate encompassing area between levels II and III) selectively isolated by incubating necropsy tissues in protease mixture	2-fold; Benjamini-Hochberg; FDR < 0.05	Exposure to formaldehyde at 2 ppm caused induction of genes involved in cellular stress responses—thiol transport/reduction, inflammation, and cell proliferation—at all exposure durations. Exposure to formaldehyde at 6 ppm or greater resulted in changes in expression of genes involved in cell-cycle regulation, DNA repair, and apoptosis.	Andersen et al. 2010
Rats: male F344/CrlBR	Formaldehyde vapor or instillation Vapor: 0, 0.7, 2, and 6 ppm 6 hours/day for 5 days/week for up to 3 weeks (5 per dose per time) Vapor: 15 ppm for 6 hours (10 exposed, 5 controls) Instillation: 400 mM x 40 µL per nostril, 6 hours (10 exposed, 5 controls)	Nasal surface epithelial cells (lateral meatus and nasoturbinate encompassing area between levels II and III) selectively isolated by incubating necropsy tissues in protease mixture	1.5-fold; Benjamini-Hochberg; FDR < 0.05	No differentially expressed genes were detected after exposure to formaldehyde vapor at 0.7 ppm. Exposure at 2 and 6 ppm resulted in up to 15 and 54 differentially expressed genes, respectively, at different timings over the course of the 3-week exposure. Exposure at 15 ppm caused 745 differentially expressed genes within 24-hour period, and exposure by instillation (400 mM x 40 µL per nostril) caused 2,553 differentially expressed genes within 24-hour period. About 75% of differentially expressed genes caused by exposure at 15 ppm were also affected by exposure via instillation, and these genes were enriched in gene ontology categories of wound response, apoptotic regulation, inflammation, and receptor tyrosine kinase signaling.	Andersen et al. 2008

(Continued)

159

160

TABLE 3-13 Continued

Model	Subjects	Exposure	Sample	Criteria[a]	Main conclusions	Reference
	Rats: male F344	Formaldehyde instillation 400 mM formaldehyde (n = 3) or water (n = 4) x 40 µL per nostril, 24 hours	Nasal epithelial cell lysis by direct instillation of Trizol reagent	Benjamini-Hochberg; FDR < 0.05 or 0.1	Exposure to formaldehyde caused differential gene expression. These genes were enriched in pathways relevant to xenobiotic metabolism, cell cycle, apoptosis, and DNA repair.	Hester et al. 2003
Cell culture	Primary culture human nasal epithelial cells (commercial product, derived from three Caucasian women)	20 or 100 µM for 2 hours; 50, 100, 200 µM for 4 hours; 100 or 200 µM for 24 hours; 20 or 50 µM for 24 hours with 4 consecutive repeats; no exposure control	Total cell lysate	2-fold; $p < 0.05$ (t test); no FDR correction	Exposure to 100 and 200 µM formaldehyde for 4 hours changed expression of 153 and 887 genes, respectively. Exposure to 50 µM formaldehyde for 24 hours with 4 repeats changed expression of 143 genes. Less than 10 differentially expressed genes were observed with all other conditions. Genes upregulated by exposure to 200 µM formaldehyde for 4 hours were enriched for apoptosis regulation and stress response.	Neuss et al. 2010b
	Human A549 lung-cancer cell line (adenocarcinoma, alveolar basal epithelial)	0 or 83.2 µM for 2 hours	Total cell lysate	1.5-fold; Benjamini-Hochberg;FDR < 0.05	Exposure to 83.2 µM formaldehyde for 2 hours caused 66 differential gene expressions, which were enriched for apoptosis regulation, transcription, and DNA damage (upregulated genes) or transcription (downregulated genes).	Cheah et al. 2013
	Human TK6 B lymphoblastoid cells	0, 50, 100, or 200 uM for 4 or 24 hours	Total cell lysate	1.5-fold and 2-fold; $p < 0.05$ (t test); FDR < 0.1 (multi-variable permutation test)	Exposure to 50 µM formaldehyde did not cause significant transcriptomal changes. Exposure to 200 µM formaldehyde caused 2,147 and 2,502 differentially expressed genes after 4 or 24 hours of exposure, respectively. Exposure to 100 µM formaldehyde for 4 hours caused 1,367 differentially expressed genes, whereas	Kuehner et al. 2013

exposure to the same concentration of formaldehyde for 24 hours caused only 2 differentially expressed genes. Genes upregulated after exposure to 200 μM formaldehyde for 24 hours were enriched for transcription, transport, protein phosphorylation, signal transduction, and apoptosis.

Abbreviation: FDR, false discovery rate.
[a]Criteria for defining differentially expressed genes.
[b]MCL1 isoform 1 is antiapoptotic, whereas isoform 2 is proapoptotic.

the blood specimens, statistically significant differential expression of 11 mRNA transcripts was observed. However, the authors concluded that these were "minor" effects that reflected assay variability and that inhalation of formaldehyde did not cause alterations in the expression of genes in either the nasal or blood samples. In the absence of appropriate negative exposure control groups, appropriate positive controls, or detailed power-analysis discussion, the committee was unable to determine whether the results of this study supported the absence of transcriptomal effects after exposure to formaldehyde or whether the study design provided sufficient discovery power in light of the small number of study subjects (six to eight per group).

Rager et al. (2013) examined maxilloturbinate necropsy specimens of nasal epithelial tissues from macaques and observed significant changes in expression of micro-RNAs after exposure to formaldehyde at 6 ppm 6 hours/day for 2 days. Using real-time quantitative polymerase chain reaction methods, the authors confirmed significant induction of miR-125b expression and concomitant suppression of its target mRNA transcripts, including proapoptotic genes *BAK1*, *CASP2*, *MAP2K7*, and *MCL1*.

Two other studies examined transcriptomal effects in nasal epithelial cells of F344 rats that were exposed to formaldehyde via vapor or instillation into the nostrils (Hester et al. 2003; Andersen et al. 2010). These studies collectively demonstrated that exposure to formaldehyde, either by inhalation (2 ppm or higher for 6 hours or longer) or by intranasal instillation (40 µL of a 400 mM solution for 6 hours or longer), resulted in significant changes in expression of the mRNA transcripts that encode proteins involved in cell-cycle regulation, DNA repair, wound response, inflammation, and regulation of apoptosis. In comparison, data obtained after exposure to lower doses of formaldehyde were mostly insignificant.

Three cell-culture experiments—one that used primary cultures of human nasal epithelial cells (Neuss et al. 2010a), one that used human A549 lung alveolar basal epithelial cancer cells (Cheah et al. 2013), and one that used human TK6 lymphoblastoid cells (Kuehner et al. 2013)—demonstrated significant formaldehyde-related changes in expression of mRNA transcripts that encode proteins involved in apoptosis regulation, stress response, transcription, DNA damage, transport, and signal transduction. Relatively high concentrations of formaldehyde—greater than 83.2 µM for 2 hours (Cheah et al. 2013) or greater than 100 µM for 4 hours (Neuss et al. 2010a; Kuehner et al. 2013)—resulted in transcriptomal changes, whereas exposure to lower concentrations of formaldehyde did not have detectable effects even after prolonged exposure.

The committee found multiple studies that reported transcriptional responses in nasal cavity epithelial cells from experimental animals exposed to formaldehyde vapor at doses of 2 ppm or greater. The transcriptomal responses were indicative of cell apoptosis, DNA damage, and proliferation, which are relevant to carcinogenesis. The committee notes that the doses are relevant to occupational human exposure to formaldehyde. The committee did not identify studies that considered the transcriptomal effects of chronic, low-dose exposure to formaldehyde in the nasal epithelial cells, peripheral blood, or any other tissues of human or animal models.

SUMMARY OF EVIDENCE

The statement of task specifically asked the committee to "integrate the level-of-evidence conclusions, and considering all relevant information in accordance with the RoC listing criteria, make an independent listing recommendation for formaldehyde and provide scientific justification for its recommendation" (Appendix B). The committee notes that the term *integrate* does not have a standard definition in the context of hazard assessment. The committee understood the term in its conventional sense of bringing together parts into a whole. To be listed as "reasonably anticipated as a human carcinogen" or "known to be a human carcinogen", the RoC listing criteria only requires information to be integrated across human studies or across animal studies, and supporting information can be derived from mechanistic studies. Mechanistic information "can be useful for evaluating whether a relevant cancer mechanism is operating in people" (NTP 2010, p. iv), but a known mechanism is not required for a substance to be listed in the RoC. In the subsections below, the committee summarizes human, experimental animal, and mechanistic information on nasopharyngeal and sinonasal cancer and myeloid leukemia. Summaries were not presented for other kinds of cancer because of a lack of strong evidence that formaldehyde exposure causes other kinds of cancer in humans.

Nasopharyngeal and Sinonasal Cancers

The committee found clear and convincing epidemiologic evidence of an association between formaldehyde exposure and nasopharyngeal cancer and sinonasal cancer in humans. On the basis of evidence of an association between nasopharyngeal cancer and exposure to formaldehyde in two strong studies—a large case–control study (Vaughan et al. 2000) and a large cohort study (Beane Freeman et al. 2013)—and other supporting studies that were judged to be moderately strong (Vaughan et al. 1986a,b; West et al. 1993; Hildesheim et al. 2001; Siew et al. 2012), the committee concludes that the relationship is causal and chance, bias, and confounding factors can be ruled out with reasonable confidence. For sinonasal cancer, there is evidence of an association based on a strong, well-conducted pooled case–control study (Luce et al. 2002) and other, corroborating studies that were judged to be moderately strong (Hayes et al. 1986; Olsen and Asnaes 1986; Vaughan et al. 1986a,b; Luce et al. 1993; Siew et al. 2012). The committee concludes that the relationship between formaldehyde and sinonasal cancer is causal and chance, bias, and confounding factors can be ruled out with reasonable confidence.

Several well-conducted studies in experimental animal models demonstrate an increase in nasal squamous-cell carcinoma after inhalation exposure to formaldehyde (Kerns et al.1983; Sellakumar et al. 1985; Monticello et al. 1996).

Two of the studies used F344 rats (Kerns et al. 1983; Monticello et al. 1996), and one used Sprague Dawley rats (Sellakumar et al. 1985). The evidence is corroborated by other rat studies (Feron et al. 1988; Soffritti et al. 1989; Woutersen et al. 1989; Kamata et al. 1997) and by a mouse study (Kerns et al. 1983). Although there are limitations in extrapolating findings on nasal tumors in rodents to nasopharyngeal and sinonasal cancer in humans, the experimental-animal evidence indicates that exposure to inhaled formaldehyde is associated with carcinogenic effects on tissues at the portal of entry.

Inhalation of formaldehyde at sufficient concentrations substantially increases formaldehyde to above the total endogenous concentration in tissues at the portal of entry in both animal and human studies. There is experimental evidence that, due to its chemical reactivity, formaldehyde exerts genotoxic and mutagenic effects and cytotoxicity followed by compensatory cell proliferation at the portal of entry[3] in animals and humans exposed to formaldehyde; this provides biologic plausibility of a relationship between formaldehyde exposure and cancer. The evidence on formaldehyde-associated DNA adducts, DNA–protein cross-links, DNA strand breaks, mutations, micronuclei, and chromosomal aberrations is consistent, strong, and specific. In addition, both temporal and exposure–response relationships have been established, most strongly in studies of rodents and nonhuman primates.

Myeloid Leukemia

The committee found clear and convincing epidemiologic evidence of an association between formaldehyde exposure and myeloid leukemia. There may also be an increase of other lymphohematopoietic cancers, although the evidence is less robust. On the basis of three strong studies with widely different coexposures (the NCI formaldehyde-industry cohort [Beane Freeman et al. 2009], the NIOSH garment-worker cohort [Meyers et al. 2013], and the NCI funeral-industry cohort [Hauptmann et al. 2009]) and several moderately strong studies (Walrath and Fraumeni 1983, 1984; Stroup et al. 1986; Coggon et al. 2014), the committee concludes that there is a causal association between formaldehyde exposure and myeloid leukemia. Chance, bias, and confounding factors can be ruled out with reasonable confidence given the consistent pattern of association in the larger studies that had good exposure assessment.

Although multiple lines of reasoning and experimental evidence indicate that it is unlikely that inhalation exposure to formaldehyde will increase formaldehyde to substantially above endogenous concentrations in tissues distant from the site of entry, there is a robust database of experimental studies of *systemic*[4]

---

[3]Defined as effects that arise from direct interaction of inhaled or ingested formaldehyde with cells or tissues.

[4]Defined as effects that occur beyond cells or tissues that have direct interaction with inhaled or ingested formaldehyde.

mechanistic events that have been observed after exposure to formaldehyde. The committee notes that it is plausible that some of the systemic effects, notably findings of genotoxicity and transcriptional changes in circulating blood cells, may have resulted from the exposure of the cells at the portal of entry (for example, lymphoid tissue in the nasal mucosa). The mechanistic events that were considered by the committee as relevant to the plausibility of formaldehyde-associated tumors beyond the portal of entry included genotoxicity and mutagenicity, hematologic effects, and effects on gene expression. Overall, in mechanistic studies of experimental animals and exposed humans, the evidence is largely consistent and strong. As shown in Table 3-10, a majority of the mammalian in vivo studies resulted in positive findings compared to negative findings (60 and 38 studies, respectively), particularly in humans (49 and 19 studies, respectively). Both temporal and exposure–response relationships have been demonstrated in studies of humans and animals exposed to formaldehyde. The committee concludes that these findings provide plausible mechanistic pathways supporting a relationship between formaldehyde exposure and cancer, even though the potential mechanisms of how formaldehyde may cause such systemic effects are not fully understood. It would be desirable to have a more complete understanding about how formaldehyde exposure may cause systemic effects, but the lack of known mechanisms should not detract from the findings of an association between formaldehyde exposure and myeloid leukemia in epidemiology studies.

The animal cancer bioassay literature provided some information relevant to myeloid leukemia. One drinking water study (Soffritti et al. 2002) reported a significant increase in lymphohematopoietic cancers following long-term exposure to formaldehyde in drinking water, but there is uncertainty regarding the finding. Of the three inhalation studies that included histopathologic examinations of non–respiratory tract tissues, two did not report leukemia (Sellakumar et al. 1985; Kamata et al. 1997). The full laboratory report (Battelle 1981) of a third study (Kerns et al. 1983) discussed findings of leukemia and lymphoma that were not found to be compound related. However, diffuse multifocal bone marrow hyperplasia in rats exposed to 15 ppm of formaldehyde for 18 months was increased in both treated males ($p = 0.0001$) and females ($p = 0.0001$). Although the Battelle finding was not a finding of malignancy, it does indicate that long-term inhaled formaldehyde may cause effects in bone marrow.

CONCLUSIONS AND LISTING RECOMMENDATION

The committee identified and evaluated relevant, publicly available, peer-reviewed literature on formaldehyde, including attention to literature published between June 10, 2011 (the release date of the substance profile for formaldehyde in the 12th RoC), and November 8, 2013. The committee applied NTP's established RoC listing criteria to the scientific evidence on formaldehyde from studies of humans, studies of experimental animals, and other studies relevant to mechanisms of carcinogenesis.

The type of information needed to meet the criteria for sufficient evidence in experimental animals is clear and transparent, as outlined in the section "Cancer Studies in Experimental Animals". In contrast, the RoC listing criteria do not provide detailed guidance about how evidence should be assembled to meet the requirement of limited evidence or sufficient evidence of carcinogenicity from studies in humans, except to note that limited evidence cannot exclude alternative explanations, such as chance, bias, or confounding factors, and to note that conclusions should be based on "scientific judgment, with consideration given to all relevant information" (NTP 2010, p. iv). In the section "Cancer Studies in Humans", the committee used scientific judgment to develop an approach to assessing the epidemiology evidence. The approach included careful review of individual studies, selection of studies that were most informative, and evaluation of informative studies on the basis of the strength, consistency, temporality, dose-response, and coherence of the evidence and on the considerations presented in Table 3-1.

The committee notes that evidence in experimental animals and a known mechanism of action is not required by the RoC listing criteria in making a listing recommendation that a substance is known to be a human carcinogen if the evidence from studies in humans is sufficient and indicates an association between exposure and human cancer. Also, and importantly, the RoC listing criteria require an association in only one type of cancer to make the determination. On the basis of the information summarized directly above for nasopharyngeal cancer, sinonasal cancer, and for myeloid leukemia, the committee makes its independent determinations as follows:

- There is sufficient evidence of carcinogenicity from studies of humans based on consistent epidemiologic findings on nasopharyngeal cancer, sinonasal cancer, and myeloid leukemia for which chance, bias, and confounding factors could be ruled out with reasonable confidence.
- There is sufficient evidence of carcinogenicity in animals based on malignant and benign tumors in multiple species, at multiple sites, by multiple routes of exposure, and to an unusual degree with regard to type of tumor.
- There is convincing relevant information that formaldehyde induces mechanistic events associated with the development of cancer in humans, specifically genotoxicity and mutagenicity, hematologic effects, and effects on gene expression.

Because there is sufficient evidence of carcinogenicity from studies in humans that indicates a causal relationship between exposure to formaldehyde and at least one type of human cancer, the committee concludes that formaldehyde should be listed in the RoC as "known to be a human carcinogen".

REFERENCES

Acheson, E.D., H.R. Barnes, M.J. Gardner, C. Osmond, B. Pannett, and C.P. Taylor. 1984. Formaldehyde in the British chemical industry. An occupational cohort study. Lancet 1(8377):611-616.

Andersen, M.E., H.J. Clewell, E. Bermudez, G.A. Willson, and R.S. Thomas. 2008. Genomic signatures and dose-dependent transitions in nasal epithelial responses to inhaled formaldehyde in the rat. Toxicol. Sci. 105(2):368-383.

Andersen, M.E., H.J. Clewell, E. Bermudez, D.E. Dodd, G.A. Willson, J.L. Campbell, and R.S. Thomas. 2010. Formaldehyde: Integrating dosimetry, cytotoxicity, and genomics to understand dose-dependent transitions for an endogenous compound. Toxicol. Sci. 118(2):716-731.

Andjelkovich, D.A., D.B. Janszen, M.H. Brown, R.B. Richardson, and F.J. Miller. 1995. Mortality of iron foundry workers. IV. Analysis of a subcohort exposed to formaldehyde. J. Occup. Environ. Med. 37(7):826-837.

Appelman, L.M., R.A. Woutersen, A. Zwart, H.E. Falke, and V.J. Feron. 1988. One-year inhalation toxicity study of formaldehyde in male rats with a damaged or undamaged nasal mucosa. J. Appl. Toxicol. 8(2):85-90.

Armstrong, R.W., P.B. Imrey, M.S. Lye, M.J. Armstrong, M.C. Yu, and S. Sani. 2000. Nasopharyngeal carcinoma in Malaysian Chinese: Occupational exposures to particles, formaldehyde and heat. Int. J. Epidemiol. 29(6):991-998.

ATSDR (Agency for Toxic Substances and Disease Registry). 1999. Toxicological Profile for Formaldehyde. U.S. Department of Health and Human Services, Public Health Service, Agency for Toxic Substances and Disease Registry, Atlanta, GA [online]. Available: http://www.atsdr.cdc.gov/toxprofiles/tp111.pdf [accessed Sept. 23, 2013].

Aydin, S., H. Canpinar, U. Undeger, D. Güc, M. Çolakoğlu, A. Kars, and N. Başaran. 2013. Assessment of immunotoxicity and genotoxicity in workers exposed to low concentrations of formaldehyde. Arch. Toxicol. 87(1):145-153.

Battelle. 1981. Final Report on a Chronic Inhalation Toxicology Study in Rats and Mice Exposed to Formaldehyde. Prepared by Battelle Columbus Laboratories, Columbus, OH, for the Chemical Industry Institute of Toxicology (CIIT), Research Triangle Park, NC. CIIT Docket No. 10922.

Beane Freeman, L.E., A. Blair, J.H. Lubin, P.A. Stewart, R.B. Hayes, R.N. Hoover, and M. Hauptmann. 2009. Mortality from lymphohematopoietic malignancies among workers in formaldehyde industries: The National Cancer Institute Cohort. J. Natl. Cancer Inst. 101(10):751-761.

Beane Freeman, L.E., A. Blair, J.H. Lubin, P.A. Stewart, R.B. Hayes, R.N. Hoover, and M. Hauptmann. 2013. Mortality from solid tumors among workers in formaldehyde industries: An update of the NCI cohort. Am. J. Ind. Med. 56(9):1015-1026.

Bertazzi, P.A., A. Pesatori, S. Guercilena, D. Consonni, and C. Zocchetti. 1989. Carcinogenic risk for resin producers exposed to formaldehyde: Extension of follow-up [in Italian]. Med. Lav. 80(2):111-122.

Blair, A., and P.A. Stewart. 1990. Correlation between different measures of occupational exposure to formaldehyde. Am. J. Epidemiol. 131(3):510-516.

Blair, A., P. Stewart, M. O'Berg, W. Gaffey, J. Walrath, J. Ward, R. Bales, S. Kaplan, and D. Cubit. 1986. Mortality among industrial workers exposed to formaldehyde. J. Natl. Cancer Inst. 76(6):1071-1084.

Blair, A., P.A. Stewart, and R.N. Hoover. 1990. Mortality from lung cancer among workers employed in formaldehyde industries. Am. J. Ind. Med. 17(6):683-699.

Blair, A., T. Zheng, A. Linos, P.A. Stewart, Y.W. Zhang, and K.P. Cantor. 2001. Occupation and leukemia: A population-based case-control study in Iowa and Minnesota. Am. J. Ind. Med. 40(1):3-14.

Bolt, H.M. 1987. Experimental toxicology of formaldehyde. J. Cancer Res. Clin. Oncol. 113(4):305-309.

Bono, R., M. Vincenti, T. Schiliro, E. Scursatone, C. Pignata, and G. Gilli. 2006. N-Methylenvaline in a group of subjects occupationally exposed to formaldehyde. Toxicol. Lett. 161(1):10-17.

Boobis, A.R., J.E. Doe, B. Heinrich-Hirsch, M.E. Meek, S. Munn, M. Ruchirawat, J. Schlatter, J. Seed, and C. Vickers. 2008. IPCS framework for analyzing the relevance of a noncancer mode of action for humans. Crit. Rev. Toxicol. 38(2):87-96.

Brondeau, M.T., P. Bonnet, J.P Guenier, P. Simon, and J. de Ceaurriz. 1990. Adrenal-dependent leucopenia after short-term exposure to various airborne irritants in rats. J. Appl. Toxicol. 10(2): 83-86.

Bucher, J.R. 2013. Follow-up Questions. Material submitted by the NAS Committee on Review of the Formaldehyde Assessment in the NTP 12th RoC and the NAS Committee on Review of the Styrene Assessment in the NTP 12th RoC, April 2, 2013.

Buss, J., K. Kuschinsky, H. Kewitz, and W. Koransky. 1964. Enteric resorption of formaldehyde [in German]. N-S Arch. Exp. Pathol. Pharmakol. 247:380-381.

Casanova, M., H.d'A. Heck, J.I. Everitt, W.W. Harrington, Jr., and J.A. Popp. 1988. Formaldehyde concentrations in the blood of rhesus monkeys after inhalation exposure. Food Chem. Toxicol. 26(8):715-716.

Casanova, M., K.T. Morgan, W.H. Steinhagen, J.I. Everitt, J.A. Popp, and H.D. Heck. 1991. Covalent binding of inhaled formaldehyde to DNA in the respiratory tract of rhesus monkey: Pharmacokinetics, rat to monkey interspecies scaling, and extrapolation to man. Fundam. Appl. Toxicol. 17(2):409-428.

Casanova-Schmitz, M., T.B. Starr, and H.D. Heck. 1984. Differentiation between metabolic incorporation and covalent binding in the labeling of macromolecules in the rat nasal mucosa and bone marrow by inhaled [14C]- and [3H]formaldehyde. Toxicol. Appl. Pharmacol. 76(1):26-44.

Cheah, N.P., J.L. Pennings, J.P. Vermeulen, F.J. van Schooten, and A. Opperhuizen. 2013. In vitro effects of aldehydes present in tobacco smoke on gene expression in human lung alveolar epithelial cells. Toxicol. In Vitro 27(3):1072-1081.

Checkoway, H., N. Pearce, and D. Kribel. 2004. Research Methods in Occupational Epidemiology, 2nd Ed. Oxford: Oxford University Press.

Checkoway, H., R.M. Ray, J.I. Lundin, G. Astrakianakis, N.S. Seixas, J.E. Camp, K.J. Wernli, E.d. Fitzgibbons, W. Li, Z. Feng, D.L. Gao, and D.B. Thomas. 2011. Lung cancer and occupational exposures other than cotton dust and endotoxin among women textile workers in Shanghai, China. Occup. Environ. Med. 68(6):425-429.

Cheng. Z., Y. Li, B. Liang, and C. Wang. 2004. Investigation of formaldehyde level and health of personnel in clinical pathology. J. Bengbu. Med. Coll. 29(3): 266-267.

Coggon, D., E.C. Harris, J. Poole, and K.T. Palmer. 2003. Extended follow-up of a cohort of British chemical workers exposed to formaldehyde. J. Natl. Cancer Inst. 95(21):1608-1615.

Coggon, D., G. Ntani, E.C. Harris, and K.T. Palmer. 2014. Upper airway cancer, myeloid leukemia, and other cancers in a cohort of British chemical workers exposed to formaldehyde. Am. J. Epidemiol. 179(11):1301-1311.

Costa, S., J. Garcia-Leston, M. Coelho, P. Coelho, C. Costa, S. Silva, B. Porto, B. Laffon, and J.P. Teixeira. 2013. Cytogenetic and immunological effects associated with occupational formaldehyde exposure. J. Toxicol. Environ. Health A 76(4-5):217-229.

Dalbey, W.E. 1982. Formaldehyde and tumors in hamster respiratory tract. Toxicology 24(1):9-14.

Dell, L., and M.J. Teta. 1995. Mortality among workers at a plastics manufacturing and research and development facility: 1946-1988. Am. J. Ind. Med. 28(3):373-384.

Doll, R., and R. Peto. 1978. Cigarette smoking and brochial carcinoma: Dose and time relationships among regular smokes and lifelong non-smokers. J. Epidemiol. Community Health 32(4):303-313.

Edling, C., B. Jarvholm, L. Andersson, and O. Axelson. 1987. Mortality and cancer incidence among workers in an abrasive manufacturing industry. Br. J. Ind. Med. 44(1):57-59.

Edrissi, B., K. Taghizadeh, and P.C. Dedon. 2013a. Quantitative analysis of histone modifications: Formaldehyde is a source of pathological N^6-formyllysine that is refractory to histone deacetylases. PLoS Genet. 9(2):e1003328.

Edrissi, B., K. Taghizadeh, B.C. Moeller, D. Kracko, M. Doyle-Eisele, J.A. Swenberg, and P.C. Dedon. 2013b. Dosimetry of N^6-formyllysine adducts following [$^{13}C^2H_2$]-formaldehyde exposures in rats. Chem. Res. Toxicol. 26(10):1421-1423.

Egle, J.L., Jr. 1972. Retention of inhaled formaldehyde, propionaldehyde, and acrolein in the dog. Arch. Environ. Health 25(2):119-124.

Elliot, L.J., L.T. Stayner, L.M. Blade, W. Helperin, and R. Keenlyside. 1987. Formaldehyde Exposure Characterization in Garment Manufacturing Plants: A Composite Summary of Three in-depth Industrial Hygiene Surveys. Division of Surveillance, Hazard Evaluations and Field Studies, National Institute for Occupational Safety and Health, Cincinnati, OH.

EPA (U.S. Environmental Protection Agency). 2005. Guidelines for Carcinogen Risk Assessment. EPA/630/P-03/001F. Risk Assessment Forum, U.S. Environmental Protection Agency, Washington, DC [online]. Available: http://www.epa.gov/raf/publicati ons/pdfs/CANCER_GUIDELINES_FINAL_3-25-05.PDF [accessed Jan. 23, 2014].

Erdei, E., J. Bobvos, M. Brozik, A. Paldy, I. Farkas, E. Vaskovi, and P. Rudnai. 2003. Indoor air pollutants and immune biomarkers among Hungarian asthmatic children. Arch. Environ. Health 58(6):337-347.

Feron, V.J., J.P. Bruyntjes, R.A. Woutersen, H.R. Immel, and L.M. Appelman. 1988. Nasal tymours in rats after short-term exposure to a cytotoxic concentration of formaldehyde. Cancer Lett. 39(1):101-111.

Fox, E.M. 1985. Urea formaldehyde foam insulation: Defusing a time-bomb. Am. J. Law Med. 11(1):81-104.

Frank, N.R., R.E. Yoder, J.D. Brain, and E. Yokoyama. 1969. SO2 (35S labeled) absorption by the nose and mouth under conditions of varying concentration and flow. Arch. Environ. Health 18(3):315-322.

Franks, S.J. 2005. A mathematical model for the absorption and metabolism of formaldehyde vapour by humans. Toxicol. Appl. Pharmacol. 206(3):309-320.

Gardner, M.J., B. Pannett, P.D. Winter, and A.M. Cruddas. 1993. A cohort study of workers exposed to formaldehyde in the British chemical industry: An update. Br. J. Ind. Med. 50(9):827-834.

Garschin, W.G., and L.M. Schabad. 1936. About atypical proliferation of the bronchial epithelium with the introduction of formalin into the lung tissue [in German]. Z. Krebsforsch. 43(1):137-145.

Gentry, P.R., J.V. Rodricks, D. Bachand, C. Van Landingham, A.M. Shipp, R.J. Albertini, and R. Irons. 2013. Formaldehyde exposure and leukemia: Critical review and reevaluation of the results from a study that is the focus for evidence of biological plausibility. Crit. Rev. Toxicol. 43(8):661-670.

Georgieva, A.V., J.S. Kimbell, and P.M. Schlosser. 2003. A distributed-parameter model for formaldehyde update and disposition in the rat nasal lining. Inhal. Toxicol. 15(14):1435-1463.

Gift, J.S., J.C. Caldwell, J. Jinot, M.V. Evans, I. Cote, and J.J. Vandenberg. 2013. Scientific considerations for evaluating cancer bioassays conducted by the Ramazzini Institute. Environ. Health Perspect, 121(11-12):1253-1263.

Glass, D.C., M.R. Sim, L. Fritschi, C.N. Gray, D.J. Jolley, and C. Gibbon. 2004. Leukemia risk and relevant benzene exposure period – Re: Followup time on risk estimates, Am. J. Ind. Med. 42:481-489, 2002 [letter]. Am. J. Ind. Med. 45(2):222-223.

Gloede, E., J.A. Cichocki, J.B. Baldino, and J.B. Morris. 2011. A validated hybrid computational fluid dynamic-physiologically based pharmacokinetic model for respiratory tract vapor absorption in hum and rat and its application to inhalation dosimetry of diacetyl. Toxicol. Sci. 123(1):231-246.

Goldstein, H.B. 1973. Textiles and the chemical industry: A marriage. J. Am. Assoc. Text. Chem. Color. 5(10):209-214.

Hall, A., J.M. Harrington, and T.C. Aw. 1991. Mortality study of British pathologists. Am. J. Ind. Med. 20(1):83-89.

Hansen, J., and J.H. Olsen. 1995. Formaldehyde and cancer morbidity among male employees in Denmark. Cancer Causes Control 6(4):354-360.

Hansen, J., and J.H. Olsen. 1996. Occupational exposure to formaldehyde and risk of cancer [in Danish]. Ugeskr. Laeger. 158(29):4191-4194.

Harrington, J.M., and D. Oakes. 1984. Mortality study of British pathologists 1974-80. Br. J. Ind. Med. 41(2):188-191.

Hauptmann, M., J.H. Lubin, P.A. Stewart, R.B. Hayes, and A. Blair. 2004. Mortality from solid cancers among workers in formaldehyde industries. Am. J. Epidemiol. 159(12):1117-1130.

Hauptmann, M., P.A. Stewart, J.H. Lubin, L.E. Beane Freeman, R.W. Hornung, R.F. Herrick, R.N Hoover, J.F. Fraumeni Jr., A. Blair, and R.B. Hayes. 2009. Mortality from lymphohematopoietic malignancies and brain cancer among embalmers exposed to formaldehyde. J. Natl. Cancer Inst. 101(24):1696-1708.

Hayes, R.B., J.W. Raatgever, A. de Bruyn, and M. Gerin. 1986. Cancer of the nasal cavity and paranasal sinuses, and formaldehyde exposure. Int. J. Cancer 37(4):487-492.

Hayes, R.B., A. Blair, P.A. Stewart, R.F. Herrick, and H. Mahar. 1990. Mortality of U.S. embalmers and funeral directors. Am. J. Ind. Med. 18(6):641-652.

Heck, H., and M. Casanova. 2004. The implausibility of leukemia induction by formaldehyde: A critical review of the biological evidence on distant-site toxicity. Regul. Toxicol. Pharmacol. 40(2):92-106.

Heck, H.A., E.L. White, and M. Casanova-Schmitz. 1982. Determination of formaldehyde in biological tissues by gas chromatography/mass spectrometry. Biomed. Mass. Spectrom. 9(8):347-353.

Heck, H.A., M. Casanova-Schmitz, P.B. Dodd, E.N. Schachter, T.J. Witek, and T. Tosun. 1985. Formaldehyde (CH_2O) concentrations in the blood of humans and Fischer-344 rats exposed to CH2O under controlled conditions. Am. Ind. Hyg. Assoc. J. 46(1):1-3.

Heck, H.A., M. Casanova, W.H. Steinhagen, J.I. Everitt, K.T. Morgan, and J.A. Popp. 1989. Formaldehyde toxicity: DNA-protein cross-linking studies in rats and non-human primates. Pp. 159-164 in Nasal Carcinogenesis in Rodents: Relevance to Human Risk, V.J. Feron, and M.C. Bosland, eds. Wageningen: Pudoc.

Hester, S.D., G.B. Benavides, L. Yoon, K.T. Morgan, F. Zou, W. Barry, and D.C. Wolf. 2003. Formaldehyde-induced gene expression in F344 rat nasal respiratory epithelium. Toxicology 187(1):13-24.

Hildesheim, A., M. Dosemeci, C.C. Chan, C.J. Chen, Y.J. Cheng, M.M. Hsu, I.H. Chen, B.F. Mittl, B. Sun, P.H. Levine, J.Y. Chen, L.A. Brinton, and C.S. Yang. 2001. Occupational exposure to wood, formaldehyde, and solvents and risk of nasopharyngeal carcinoma. Cancer Epidemiol. Biomarkers Prev. 10(11):1145-1153.

Hill, A.B. 1965. The environment and disease: Association or causation? Proc. R. Soc. Med. 58(5):295-300.

Holmström, M., B. Wilhelmosson, and H. Hellquist. 1989. Histological changes in the nasal mucosa in rats after long-term exposure to formaldehyde and wood dust. Acta Otolaryngol. 108(3-4):274-283.

Horton, A.W., R. Tye, and K.L. Stemmer. 1963. Experimental carcinogenesis of the lung. Inhalation of gaseous formaldehyde or an aerosol coal tar by C3H mice. J. Natl. Cancer Inst. 30:31-43.

Hosgood, H.D., III, L. Zhang, X. Tang, R. Vermeulen, Z. Hao, M. Shen, C. Qiu, Y. Ge, M. Hua, Z. Ji, S. Li, J. Xiong, B. Reiss, S. Liu, K.X. Xin, M. Azuma, Y. Xie, L. Beane Freeman, X. Ruan, W. Guo, N. Galvan, A. Blair, L. Li, H. Huang, M.T. Smith, N. Rothman, and Q. Lan. 2013. Occupational exposure to formaldehyde and alterations in lymphocyte subsets. Am. J. Ind. Med. 56(2):252-257.

IARC (International Agency for Research on Cancer). 1982. Chemicals, Industrial Processes and Industries Associated with Cancer in Humans: An Updating of IARC Monographs Volumes 1 to 29. IARC Monographs on the Evaluation of the Carcinogenic Risks to Humans Supplement 4. Lyon, France: IARC [online]. Available: http://monographs.iarc.fr/ENG/Monographs/suppl4/Suppl4.pdf [accessed June 10, 2013].

IARC (International Agency for Research on Cancer). 1995. Wood Dust and Formaldehyde. IARC Monographs on the Evaluation of the Carcinogenic Risks to Humans Vol. 62. Lyon, France: IARC [online]. Available: http://monographs.iarc.fr/ENG/Monographs/vol62/mono62.pdf [accessed June 10, 2013].

IARC (International Agency for Research on Cancer). 2006a. Formaldehyde. Pp. 39-325 in Formaldehyde, 2-Butoxyethanol and 1-tert-Butoxypropan-2-ol. IARC Monographs on the Evaluation of the Carcinogenic Risks to Humans Vol. 88. Lyon, France: IARC [online]. Available: http://monographs.iarc.fr/ENG/Monographs/vol88/mono88.pdf [June 10, 2013].

IARC (International Agency for Research on Cancer). 2006b. IARC Monographs on the Evaluation of the Carcinogenic Risks to Humans: Preamble. Lyon, France: IARC [online]. Available: http://monographs.iarc.fr/ENG/Preamble/CurrentPreamble.pdf [accessed June 10, 2013].

Jakab, M.G., T. Klupp, K. Besenyei, A. Biro, J. Major, and A. Tompa. 2010. Formaldehyde-induced chromosomal aberrations and apoptosis in peripheral blood lymphocytes of personnel working in pathology departments. Mutat. Res. 698(1-2):11-17.

Ji, Z., X. Li, M. Fromowitz, E. Mutter-Rottmayer, J. Tung, M. Smith, and L. Zhang. 2013. Formaldehyde induces micronuclei in mouse erythropoietic cells and suppresses the expansion of human erythroid progenitor cells. Toxicol. Lett. 224(2):233-239.

Jiang, S., L. Yu, J. Cheng, S. Leng, Y. Dai, Y. Zhang, Y. Niu, H. Yan, W. Qu, C. Zhang, K. Zhang, R. Yang, L. Zhou, and Y. Zheng. 2010. Genomic damages in peripheral blood lymphocytes and association with polymorphisms of three glutathione S-transferases in workers exposed to formaldehyde. Mutat. Res. 695(1-2):9-15.

Kamata, E., M. Nakadate, O. Uchida, Y. Ogawa, S. Suzuki, T. Kaneko, M. Saito, and Y. Jurokawa. 1997. Results of a 28-month chronic inhalation toxicity study of formaldehyde in male Fisher-344 rats. J. Toxicol. Sci. 22(3):239-254.

Katsnelson, B.A., T.D. Degtyareva, L.I. Privalova, I.A. Minigaliyeva, T.V. Slyshkina, V.V. Ryzhow, and O.Y. Beresneva. 2013. Attenuation of subchronic formaldehyde inhalation toxicity with oral administration of flutamate, glycine and methione. Toxicol. Lett. 220(2):181-186.

Kerns, W.D., K.L. Pavkov, D.J. Donofrio, E.J. Gralla, and J.A. Swenberg. 1983. Carcinogenicity of formaldehyde in rats and mice after long-term inhalation exposure. Cancer Res. 43(9):4382-4392.

Kim, E.M., H.Y. Lee, E.H. Lee, K.M. Lee, M. Park, K.Y. Ji, J.H. Jang, Y.H. Jeong, K.H. Lee, I.J. Yoon, S.M. Kim, M.J. Jeong, K.D. Kim, and H.S. Kang. 2013. Formaldehyde exposure impairs the function and differentiation of NK cells. Toxicol. Lett. 223(2):154-161.

Kimbell, J.S. 2006. Nasal dosimetry of inhaled gases and particles: Where do inhaled agents go in the nose? Toxicol. Pathol. 34(3):270–273.

Kimbell, J.S., R.P. Subramaniam, E.A. Gross, P.M. Schlosser, and K.T. Morgan. 2001. Dosimetry modeling of inhaled formaldehyde: Comparisons of local flux predictions in the rat, monkey, and human nasal passages. Toxicol. Sci. 64(1):100-110.

Kuehner, S., K. Holzmann, and G. Speit. 2013. Characterization of formaldehyde's genotoxic mode of action by gene expression analysis in TK6 cells. Arch. Toxicol. 87(11):1999-2012.

Kuo, H., G. Jian, C. Chen, C. Liu, and J. Lai. 1997. White blood cell count as an indicator of formaldehyde exposure. Bull. Environ. Contam. Toxicol. 59(2):261–267.

Ladeira, C., S. Viegas, E. Carolino, J. Prista, M.C. Gomes, and M. Brito. 2011. Genotoxicity biomarkers in occupational exposure to formaldehyde--the case of histopathology laboratories. Mutat. Res. 721(1):15-20.

Levine, R.J., D.A. Andjelkovich, and L.K. Shaw. 1984. The mortality of Ontario undertakers and a review of formaldehyde-related mortality studies. J. Occup. Med. 26(10):740-746.

Li, Q., Q. Mei, T. Huyan, L. Xie, S. Che, H. Yang, M. Zhang., and Q. Huang. 2013. Effects of formaldehyde exposure on human NK cells in vitro. Environ. Toxicol. Pharmacol. 36(3): 948-955.

Li, W., R.M. Ray, D.L. Gao, E.D. Fitzgibbons, N.S. Seixas, J.E. Camp, K.J. Wernli, G. Astrakianakis, Z. Feng, D.B. Thomas, and H. Checkoway. 2006. Occupational risk factors for nasopharyngeal cancer among female textile workers in Shanghai, China. Occup. Environ. Med. 63(1):39-44.

Lin, D., Y. Guo, J. Yi, D. Kuang, X. Li, H. Deng, K. Huang, L. Guan, Y. He, X. Zhang, D. Hu, Z. Zhang, H. Zheng, X. Zhang, C.M. McHale, L. Zhang, and T. Wu. 2013. Occupational exposure to formaldehyde and genetic damage in the peripheral blood lymphocytes of plywood workers. J. Occup. Health 55(4):284-291.

Lino-dos-Santos-Franco, A., M. Correa-Costa, A.C. Durao, A.P. de Oliveira, A.C. Breithaupt-Faloppa, J. de Almeida Bertoni, R.M. Oliveira-Filho, N.O. Camara, T. Marcourakis, and W. Tavares-de-Lima. 2011. Formaldehyde induces lung inflammation by an oxidant and antioxidant enzymes mediated mechanism in the lung tissue. Toxicol. Lett. 207(3):278–285.

Lino-dos-Santos-Franco, A., J.A. Gimenes-Júnior, A.P. Ligeiro-de-Oliveira, A.C. Breithaupt-Faloppa, B.G. Acceturi, L.B. Vitoretti, I.D. Machado, R.M. Oliveira-Filho, S.H. Farsky, H.T. Moriya, and T. Tavares-de-Lima. 2013. Formaldehyde inhalation reduces respiratory mechanics in a rat model with allergic lung inflammation by altering the nitric oxide/cyclooxygenase-derived products relationship. Food Chem. Toxicol. 59:731-738.

Lu, K., L.B. Collins, H. Ru, E. Bermudez, and J.A. Swenberg. 2010a. Distribution of DNA adducts caused by inhaled formaldehyde is consistent with induction of nasal carcinoma but not leukemia. Toxicol. Sci. 116(2):441-451.

Lu, K., W.J. Ye, L. Zhou, L.B. Collins, X. Chen, A. Gold, L.M. Ball, and J.A. Swenberg. 2010b. Structural characterization of formaldehyde-induced cross-links between amino acids and deoxynucleosides and their oligomers. J. Am. Chem. Soc. 132(10):3388-3399.

Lu, K., B. Moeller, M. Doyle-Eisele, J. McDonald, J.A. Swenberg. 2011. Molecular dosimetry of N2-hydroxymethyl-dG DNA adducts in rats exposed to formaldehyde. Chem. Res. Toxicol. 24(2):159-161.

Luce, D., M. Gerin, A. Leclerc, J.F. Morcet, J. Brugere, and M. Goldberg. 1993. Sinonasal cancer and occupational exposure to formaldehyde and other substances. Int. J. Cancer. 53(2):224-231.

Luce, D., A. Leclerc, D. Begin, P.A. Demers, M. Gerin, E. Orlowski, M. Kogevinas, S. Belli, I. Bugel, U. Bolm-Audorff, L.A., Brinton, P. Comba, L. Hardell, R.B. Hayes, C. Magnani, E. Merler, S. Preston-Martin, T.L. Vaughan, W. Zheng, and P. Boffetta. 2002. Sinonasal cancer and occupational exposures: A pooled analysis of 12 case-control studies. Cancer Causes Control 13(2):147-157.

Luo, J., M. Hendryx, and A. Ducatman. 2011. Association between six environmental chemicals and lung cancer incidence in the United States. J. Environ. Public Health (2011): Art. 463701.

Lyapina, M., G. Zhelezova, E. Petrova, and M. Boev. 2004. Flow cytometric determination of neutrophil respiratory burst activity in workers exposed to formaldehyde. Int. Arch. Occup. Environ. Health 77(5):335-340.

Madison, R.E., A. Broughton, and J.D. Thrasher. 1991. Immunologic biomarkers associated with an acute exposure to exothermic byproducts of a ureaformaldehyde spill. Environ. Health Perspect. 94: 219-223.

Mahboubi, A., A. Koushik, J. Siemiatycki, J. Lavoué, and M.C. Rousseau. 2013. Assessment of the effect of occupational exposure to formaldehyde on the risk of lung cancer in two Canadian population-based case-control studies. Scand. J. Work Environ. Health 39(4):401-410.

Malarkey, D.E., and J.R. Bucher. 2011. Summary Report of the National Toxicology Program and Environmental Protection Agency-Sponsored Review of Pathology Materials from Selected Ramazzini Institute Rodent Cancer Bioassays, November 29, 2011. National Toxicology Program [online]. Available: http://ntp.niehs.nih.gov/NTP/About_NTP/Partnerships/International/SummaryPWG_Report_RI_Bioassays.pdf [accessed Mar. 12, 2014].

McGwin, G., J. Lienert, and J.L. Kennedy, Jr. 2011. Formaldehyde exposure and asthma in children: A systematic review. Cien. Saude Colet. 16(9):3845-3851.

Meyers, A.R., L.E. Pinkerton, and M.J. Hein. 2013. Cohort mortality study of garment industry workers exposed to formaldehyde: Update and internal comparisons. Am. J. Ind. Med. 56(9):1027-1039.

Moeller, B.C., K. Lu, M. Doyle-Eisele, J. McDonald, A. Gigliotti, and J.A. Swenberg. 2011. Determination of N2-hydroxymethyl-dG adducts in the nasal epithelium and bone marrow of nonhuman primates following 13CD2-formaldehyde inhalation exposure. Chem. Res. Toxicol. 24(2):162-164.

Monticello, T.M., K.T. Morgan, J.I. Everitt, and J.A. Popp. 1989. Effects of formaldehyde gas on the respiratory tract of rhesus monkeys. Pathology and cell proliferation. Am. J. Pathol. 134(3):515-527.

Monticello, T.M., J.A. Swenberg, E.A. Gross, J.R. Leininger, J.S. Kimbell, S. Seilkop, T.B. Starr, J.E. Gibson, and K.T. Morgan. 1996. Correlation of regional and non-linear formaldehyde-induced nasal cancer with proliferating populations of cells. Cancer Res. 56(5):1012-1022.

Muller, P., G. Raabe, and D. Schumann. 1978. Leukoplakia induced by repeated deposition of folmalin in rabbit oral mucosa. Long –term experiments with a new 'oral tank". Exp. Pathol. 16(1-6):36-42.

Neuss, S., K. Holzmann, and G. Speit. 2010a. Gene expression changes in primary human nasal epithelial cells exposed to formaldehyde in vitro. Toxicol. Lett. 198(2):289-295.

Neuss, S., B. Moepps, and G. Speit. 2010b. Exposure of human nasal epithelial cells to formaldehyde does not lead to DNA damage in lymphocytes after co-cultivation. Mutagenesis 25(4):359-364.

Nielsen, G.D., S.T. Larsen, and P. Wolkoff. 2013. Recent trend in risk assessment of formaldehyde exposures from indoor air. Arch. Toxicol. 87(1):73-98.

NRC (National Research Council). 2011. Review of the Environmental Protection Agency's Draft IRIS Assessment of Formaldehyde. Washington, DC: National Academies of Science.

NRC (National Research Council). 2014. Review of EPA's Integrated Risk Information System (IRIS) Process. Washington, DC: The National Academies Press.

NTP (National Toxicology Program). 2010. Report on Carcinogens Background Document for Formaldehyde, January 22, 2010. U.S. Department of Health and Human Services, Public Health Service, National Toxicology Program, Research Triangle Park, NC [online]. Available: http://ntp.niehs.nih.gov/ntp/roc/twelfth/2009/November/Formaldehyde_BD_Final.pdf [accessed July 17, 2013].

NTP (National Toxicology Program). 2011. Formaldehyde. Pp. 195-205 in Report on Carcinogens, 12th Ed. U.S. Department of Health and Human Services, Public Health Service, National Toxicology Program, Research Triangle Park, NC [online]. Available: http://ntp.niehs.nih.gov/ntp/roc/twelfth/profiles/formaldehyde.pdf [accessed July 17, 2013].

Olsen, J.H., and S. Asnaes. 1986. Formaldehyde and the risk of squamous cell carcinoma of the sinonasal cavities. Br. J. Ind. Med. 43(11):769-774.

Olsen, J.H., S.P. Jensen, M. Hink, K. Faurbo, N.O. Breum, and O.M. Jensen. 1984. Occupational formaldehyde exposure and increased nasal cancer risk in man. Int. J. Cancer 34(5):639-644.

Ott, M.G., M.J. Teta, and H.L. Greenberg. 1989. Lymphatic and hematopoietic tissue cancer in a chemical manufacturing environment. Am. J. Ind. Med. 16(6):631-643.

Pala, M., D. Ugolini, M. Ceppi, F. Rizzo, L. Maiorana, C. Bolognesi, T. Schiliro, G. Gilli, P. Bigatti, R. Bono, and D. Vecchio. 2008. Occupational exposure to formaldehyde and biological monitoring of Research Institute workers. Cancer Detect. Prev. 32(2):121-126.

Partanen, T., T. Kauppinen, R. Luukkonen, T. Hakulinen, and E. Pukkala. 1993. Malignant lymphomas and leukemias, and exposures in the wood industry: An industry-based case-referent study. Int. Arch. Occup. Environ. Health 64(8):593-596.

Patterson, D.L., E.A. Gross, M.S. Bogdanffy, and K.T. Morgan. 1986. Retention of formaldehyde gas by the nasal passages of F344 rats. Toxicologist 6:55.

Paustenbach, D., Y. Alarie, T. Kulle, N. Schachter, R. Smith, J. Swenberg, H. Witschi, and S.B. Horowitz. 1997. A recommended occupational exposure limit for formaldehyde based on irritation. J. Toxicol. Environ. Health 50(3):217-263.

Pesch, B., C.B. Pierl, M. Gebel, I. Gross, D. Becker, G. Johnen, H.P. Rihs, K. Donhuijsen, V. Lepentsiotis, M. Meier, J. Schulze, and T. Bruning. 2008. Occupational risks for adenocarcinoma of the nasal cavity and paranasal sinuses in the German wood industry. Occup. Environ. Med. 65(3):191-196.

Peto, J., H. Seidman, and I.J. Selikoff. 1982. Mesothelioma mortality in asbestos workers: Implications for models of carcinogenesis and risk assessment. Br. J. Cancer 45(1):124-135.

Pinkerton, L.E., M.J. Hein, and L.T. Stayner. 2004. Mortality among a cohort of garment workers exposed to formaldehyde: An update. Occup. Environ. Med. 61(3):193-200.

Pongsavee, M. 2011. In vitro study of lymphocyte antiproliferation and cytogenetic effect by occupational formaldehyde exposure. Toxicol. Ind. Health 27(8):719-723.

Qian, R.J., P.H. Zhang, T.L. Duang, and N.L. Yao. 1988. Investigation on occupational hazards of formaldehyde exposure [abstract]. Ind. Hyg. Occup. Dis. 14(2):101.

Rager, J.E., B.C. Moeller, M. Doyle-Eisele, D. Kracko, J.A. Swenberg, and R.C. Fry. 2013. Formaldehyde and epigenetic alterations: MicroRNA changes in the nasal epithelium of nonhuman primates. Environ. Health Perspect. 121(3):339-344.

Ren, X., Z. Ji, C.M. McHale, J. Yuh, J. Bersonda, M. Tang, M.T. Smith, and L. Zhang. 2013. The impact of FANCD2 deficiency on formaldehyde-induced toxicity in human lymphoblastoid cell lines. Arch. Toxicol. 87(1):189-196.

Richardson, D.B. 2009. Multistate modeling of leukemia in benzene workers: A simple approach to fitting the 2-stage clonal expansion model. Am. J. Epidemol. 169(1):78-85.

Richardson, D.B., C. Terschuren, and W. Hoffmann. 2008. Occupational risk factors for non-Hodgkin's lymphoma: A population-based case-control study in Northern Germany. Am. J. Ind. Med. 51(4):258-268.

Rinsky, R.A., A.B. Smith, R. Hornung, T.G. Filloon, R.J. Young, A.H. Okun, and P.J. Landrigan. 1987. Benzene and leukemia: An epidemiologic risk assessment. N Engl. J. Med. 316(17):1044-1050.

Roush, G.C., J. Walrath, L.T. Stayner, S.A. Kaplan, J.T. Flannery, and A. Blair. 1987. Nasopharyngeal cancer, sinonasal cancer, and occupations related to formaldehyde: A case-control study. J. Natl. Cancer Inst. 79(6):1221-1224.

Rusch, G.M., J.J. Clary, W.E. Rinehart, and H.F. Bolte. 1983. A 26-week inhalation toxicity study with formaldehyde in the monkey, rat, and hamster. Toxicol. Appl. Pharmacol. 68(3):329-343.

Sasaki, Y., T. Ohtani, Y. Ito, M. Mizuashi, S. Nakagawa, T. Furukawa, A. Horii, and S. Aiba. 2009. Molecular events in human T cells treated with diesel exhaust particles or formaldehyde that underline their diminished interferon-gamma and interleukin-10 production. Int. Arch. Allergy Immunol. 148(3):239-250.

Schmid, O., and G. Speit. 2007. Genotoxic effects induced by formaldehyde in human blood and implications for the interpretation of biomonitoring studies. Mutagenesis 22(1):69-74.

Schroeter, J.D., J.S. Kimbell, E.A. Gross, G.A. Willson, D.C. Dorman, Y.M. Tan, and H.J. Clewell III. 2008. Application of physiological computational fluid dynamics models to predict interspecies nasal dosimetry of inhaled acrolein. Inhal. Toxicol. 20(3):227-243.

Sellakumar, A.R., C.A. Snyder, J.J. Solomon, and R.E. Albert. 1985. Carcinogenicity of formaldehyde and hydrogen chloride in rats. Toxicol. Appl. Pharmacol. 81(3 Pt 1):401-406.

She, Y., Y. Li, Y. Liu, G. Asai, S. Sun, J. He, Z. Pan, and Y. Cui. 2013. Formaldehyde induces toxic effects and regulates the expression of damage response genes in BM-MSCs. Acta Biochim. Biophys. Sin. 45(12):1011-1020.

Siew, S.S., T. Kauppinen, P. Kyyronen, P. Heikkila, and E. Pukkala. 2012. Occupational exposure to wood dust and formaldehyde and risk of nasal, nasopharyngeal, and lung cancer among Finnish men. Cancer. Manag. Res. 4:223-232.

Silver, S.R., R.A. Rinsky, S.P. Copper, R.W. Homung, and D. Lai. 2002. Effects of follow-up time on risk estimates: A longitudinal examination of the relative risks of leukemia and multiple myeloma in a rubber hydrochloride cohort. Am. J. Ind. Med. 42(6):481-489.Smith, T.J., and D. Kriebel. 2010. A Biologic Approach to Environmental Assessment and Epidemiology. New York: Oxford University Press.

Smith, T.J., and D. Kriebel. 2010. A Biologic Approach to Environmental Assessment and Epidemiology. New York: Oxford University Press.

Soffritti, M., C. Maltoni, F. Maffei, and R. Biagi. 1989. Formaldehyde: An experimental multipotential carcinogen. Toxicol. Ind. Health 5(5):699-730.

Soffritti, M. F. Belpoggi, L. Lambertin, M. Lauriola, M. Padovani, and C. Maltoni. 2002. Results of long-term exposreimental studies on the carcinogeneicity of formaldehyde and acetaldehyde in rats. Ann. N.Y. Acad. Sci. 982:87-105.

Souza, A., and R. Devi. 2014. Cytokinesis blocked micronucleus assay of peripheral lymphocytes revealing the genotoxic effect of formaldehyde exposure. Clin. Anat. 27(3):308-312.

Speit, G., P. Schutz, J. Hogel, and O. Schmid. 2007. Characterization of the genotoxic potential of formaldehyde in V79 cells. Mutagenesis 22(6):387-394.

Speit, G., S. Neuss, and O. Schmid. 2010. The human lung cell line A549 does not develop adaptive protection against the DNA-damaging action of formaldehyde. Environ. Mol. Mutagen. 51(2):130-137.

Stayner, L., A.B. Smith, G. Reeve, L. Blade, L. Elliott, R. Keenlyside, and W. Halperin. 1985. Proportionate mortality study of workers in the garment industry exposed to formaldehyde. Am. J. Ind. Med. 7(3):229-240.

Stayner, L.T., L. Elliott, L. Blade, L., R. Keenlyside, and W. Halperin. 1988. A retrospective cohort mortality study of workers exposed to formaldehyde in the garmet industry. Am. J. Ind. Med. 13(6):667-681.

Stellman, S.D., P.A. Demers, D. Colin, and P. Boffetta. 1998. Cancer mortality and wood dust exposure among participants in the American Cancer Society Cancer Prevention Study II (CPS-II). Am. J. Ind. Med. 34(3):229-237.

Stern, F., J. Beaumont, W. Halperin, L. Murthy, B. Hills, and J. Fajen. 1987. Mortality of chrome leather tannery workers and chemical exposure in tanneries. Scand. J. Work Environ. Health 13(2):108-117.

Stern, F.B. 2003. Mortality among chrome leather tannery workers: An update. Am. J. Ind. Med. 44(2):197-206.

Stewart, P., A. Blair, D. Cubit, R. Bales, S. Kaplan, J. Ward, W. Gaffey, M. O'Berg, and J. Walrath. 1986. Estimating historical exposures to formaldehyde in a retrospective mortality study. Appl. Ind. Hyg. 1(1):34-41.

Stewart, P.A., R.F. Herrick, C.E. Feigley, D.F. Utterback, R. Hornung, H. Mahar, R. Hayes, D.E. Douthit, and A. Blair. 1992. Study design for assessing exposures of embalmers for a case-control study. Part I. Monitoring results. Appl. Occup. Environ. Hyg. 7(8):532-540.

Stroup, N.E., A. Blair, and G.E. Erikson. 1986. Brain cancer and other causes of death in anatomists. J. Natl. Cancer Inst. 77(6):1217-1224.
Swenberg, J.A., B.C. Moeller, K. Lu, J.E. Rager, R.C. Fry, and T.B. Starr. 2013. Formaldehyde carcinogenicity research: 30 years and counting for mode of action, epidemiology, and cancer risk assessment. Toxicol. Pathol. 41(2):181-189.
Takahashi, M., R. Hasegawa, F. Furukawa, K. Toyoda, H. Sato, and Y. Hayashi. 1986. Effects of ethanol, potassium metabisulfite, formaldehyde and hydrogen peroxide on gastric carcinogenesis in rats after initiation with N-methyl-N'-nitro-N-nitrosoguanidine. Jpn J. Cancer Res. 77(2):118-124.
Tang, L.X., and Y.S. Zhang. 2003. Health investigation on workers exposed to formaldehyde. Occup. Health 19(7):34-35.
Tang, X., Y. Bai, A. Duong, M.T. Smith, L. Li, and L. Zhang. 2009. Formaldehyde in China: Production, consumption, exposure levels, and health effects. Environ. Int. 35(8):1210-1224.
Thomas, D.C. 2009. Some special-purpuse design. Pp. 92-109 in Statistical Methods in Environmental Epidemiology, 1st Ed. Oxford: Oxford University Press.
Thompson, C.M., R.P. Subramanian, and R.C. Grafstrom. 2008. Mechanistic and dose considerations for supporting adverse pulmonary physiology in response to formaldehyde. Toxicol. Appl. Pharmacol. 233(3):355-359.
Til, H.P., R.A. Woutersen, V.J. Feron, V.H. Hollanders, H.E. Falke, and J.J. Clary. 1989. Two-year drinking-water study of formaldehyde in rats. Food Chem. Toxicol. 27(2):77-87.
Tobe, M., K. Naito, and Y. Kurokawa. 1989. Chronic toxicity study on formaldehyde administered orally to rats. Toxicology 56(1):79-86.
Tong, Z.M., S.X. Zhu, and J. Shi. 2007. Effect of formaldehyde on blood component and blood biochemistry of exposed workers. Chin. J. Ind. Med. 20(6):409-410.
Vargova, M., S. Janota, J. Karelova, M. Barancokova, and M. Sulcova. 1992. Analysis of the health risk of occupational exposure to formaldehyde using biological markers. Analusis 20(8):451-454.
Vaughan, T.L., C. Strader, S. Davis, and J.R. Daling. 1986a. Formaldehyde and cancers of the pharynx, sinus and nasal cavity: I. Occupational exposures. Int. J. Cancer 38(5):677-683.
Vaughan, T.L., C. Strader, S. Davis, and J.R. Daling. 1986b. Formaldehyde and cancers of the pharynx, sinus and nasal cavity: II. Residential exposures. Int. J. Cancer 38(5):685-688.
Vaughan, T.L., P.A. Stewart, K. Teschke, C.F. Lynch, G.M. Swanson, J.L. Lyon, and M. Berwick. 2000. Occupational exposure to formaldehyde and wood dust and nasopharyngeal carcinoma. Occup. Environ. Med. 57(6):376-384.
Viegas, S., C. Ladeira, C. Nunes, J. Malta-Vacas, M. Gomes, M. Brito, P. Mendonca, and J. Prista. 2010. Genotoxic effects in occupational exposure to formaldehyde: A study in anatomy and pathology laboratories and formaldehyde-resins production. J. Occup. Med. Toxicol. 5(1):25.
Walrath, J., and J.F. Fraumeni, Jr. 1983. Mortality patterns among embalmers. Int. J. Cancer 31(4):407-411.
Walrath, J., and J.F. Fraumeni, Jr. 1984. Cancer and other causes of death among embalmers. Cancer Res. 44(10):4638-4641.
Watanabe, F., T. Matsunaga, T. Soejima, and Y. Iwata. 1954. Study of the carcinogenicity of aldehyde. 1. Experimentally produced rat sarcomas by repeated injections of aqueous solution of formaldehyde [in Japanese]. Gan. 45(2-3):451-452.

West, S., A. Hildesheim, and M. Dosemerci. 1993. Non-viral risk factors for nasopharyngeal carcinoma in the Philippines: Results from a case-control study. Int. J. Cancer 55(5):722-727.

Wilmer, J.W., R.A. Woutersen, L.M. Appelman, W.K. Leeman, and V.J. Feron. 1989. Subchronic (13-week) inhalation toxicity study of formaldehyde in male rats: 8-hour intermittent versus 8-hour continuous exposures. Toxicol. Lett. 47(3):287-293.

Woutersen, R.A., L.M. Appelman, J.W. Wilmer, H.E. Falke, and V.J. Feron. 1987. Subchronic (13-week) inhalation toxicity study of formaldehyde in rats. J. Appl. Toxicol. 7(1):43-49.

Woutersen, R.A., A. van Garderen-Hoetmer, J.P. Bruijntjes, A. Zwart, and V.J. Feron. 1989. Nasal tumors in rats after severe injury to the nasal mucosa and prolonged exposure to 10ppm formaldehyde. J. Appl. Toxicol. 9(1):39-46.

Wu, Y., H. You, P. Ma, L. Li, Y. Yuan, J. Li, X.Ye, X. Liu, H. Yao, R. Chen, K. Lai, and X. Yang. 2013. Role of transient receptor potential ion channels and evoked levels of neuropeptides in a formaldehyde-induced model of asthma in BALB/c mice. PLoS One 8(5):e62827.

Yang, W.H. 2007. Hemogram of workers exposed to low concentration of formaldehyde. Pract. Prev. Med. 14(3):792-799.

Ye, X., W. Yan, H. Xie, M. Zhao, and C. Ying. 2005. Cytogenetic analysis of nasal mucosa cells and lymphocytes from high-level long-term formaldehyde exposed workers and low-level short-term exposed waiters. Mutat. Res. 588(1):22-27.

Ye, X., Z. Ji, C. Wei, C. McHale, S. Ding, R. Thomas, X. Yang, and L. Zhang. 2013. Inhaled formaldehyde induces DNA-protein crosslinks and oxidative stress in bone marrow and other distant organs of exposed mice. Environ. Mol. Mutagen. 54(9):705-718.

Ying, C.J., X.L. Ye, H. Xie, W.S. Yan, M.Y. Zhao, T. Xia, and S.Y. Yin. 1999. Lymphocyte subsets and sister-chromatid exchanges in the students exposed to formaldehyde vapor. Biomed. Environ. Sci. 12(2):88-94.

Zeka, A., R. Gore, and D. Kriebel. 2011. The two-stage clonal expansion model in occupational cancer epidemiology: Results from three cohort studies. Occup. Environ. Med. 68(8):618-624.

Zeller, J., S. Neuss, J.U. Mueller, S. Kühner, K. Holzmann, J. Högel, C. Klingmann, T. Bruckner, G. Triebig, and G. Speit. 2011a. Assessment of genotoxic effects and changes in gene expression in humans exposed to formaldehyde by inhalation under controlled conditions. Mutagenesis 26(4):555-561.

Zeller, J., A. Ulrich, J.U. Mueller, C. Riegert, S. Neuss, T. Bruckner, G. Triebig, and G. Speit. 2011b. Is individual nasal sensitivity related to cellular metabolism of formaldehyde and susceptibility towards formaldehyde-induced genotoxicity? Mutat. Res. 723(1):11-17.

Zhang, L.P., X.J. Tang, N. Rothman, R. Vermeulen, Z. Ji, M. Shen, C. Qiu, W. Guo, S. Liu, B. Reiss, L.B. Freeman, Y. Ge, A.E. Hubbard, M. Hua, A. Blair, N. Galvan, X. Ruan, B.P. Alter, K.X. Xin, S. Li, L.E. Moore, S. Kim, Y. Xie, R.B. Hayes, M. Azuma, M. Hauptmann, J. Xiong, P. Stewart, L. Li, S.M. Rappaport, H. Huang, J.F. Fraumeni, Jr., M.T. Smith, and Q. Lan. 2010. Occupational exposure to formaldehyde, hematotoxicity, and leukemia-specific chromosome changes in cultured myeloid progenitor cells. Cancer Epidemiol. Biomarkers Prev. 19(1):80-88.

Zhang, Y., X. Liu, C. McHale, R. Li, L. Zhang, Y. Wu, X. Ye, X. Yang, and S. Ding. 2013. Bone marrow injury induced via oxidative stress in mice by inhalation exposure to formaldehyde. PLoS One 8(9):e74974.

Appendix A

Biographic Information on the Committee to Review the Formaldehyde Assessment in the National Toxicology Program 12th Report on Carcinogens

Alfred O. Berg *(Chair)* is professor emeritus at the University of Washington. His previous research interests are centered around evidence-based research and policy in family medicine, including the development and use of clinical practice guidelines as practice and teaching tools. Dr. Berg has served on many national expert panels using evidence-based methods to guide practice and policy; he has been chair of the US Preventive Services Task Force, cochair of the Otitis Media Panel convened by the Agency for Health Care Policy and Research, chair of the US Centers for Disease Control and Prevention (CDC) Sexually Transmitted Disease Treatment Guidelines Panel, and member of the American Medical Association–CDC panel producing guidelines for adolescent preventive services. He was the founding chair of the CDC Panel on Evaluation of Genomic Applications in Practice and Prevention and chair of the National Institutes of Health State-of-the-Science Conference on Family History and Improving Health. Dr. Berg is a member of the Institute of Medicine and has served as a member and chair of several National Academies committees. He is currently a member of the Committee on Governance and Financing of Graduate Medical Education and of the Committee on the Assessment of Studies of Health Outcomes Related to the Recommended Childhood Immunization Schedule. Dr. Berg earned an MD from Washington University in St. Louis, Missouri.

John C. Bailar III is an emeritus professor of the University of Chicago. His research interests have included trends in cancer; assessing health risks, such as the risks associated with new chemicals; and misconduct in science. His expertise includes statistics, biostatistics, epidemiology, and environmental and occupational hazards. Dr. Bailar worked at the National Cancer Institute for 22 years, and he has held academic appointments at Harvard University, McGill University, and the University of Chicago. For 11 years, he was the statistical consultant and a member of the editorial board for *The New England Journal of Medicine*.

He was a MacArthur Fellow from 1990 to 1995 and was elected to both the Institute of Medicine and the International Statistical Institute. Dr. Bailar has served as a member and as chair of many National Academies committees. His most recent committee work has included participation in the Committee on the Analysis of Cancer Risks in Populations Near Nuclear Facilities—Phase I and the Committee to Review Possible Toxic Effects from Past Environmental Containment at Fort Detrick. Dr. Bailar earned an MD from Yale and a PhD in statistics from American University.

A. Jay Gandolfi recently completed a position as associate dean for research and graduate studies at the University of Arizona and is now a professor emeritus. His research interests include the molecular and cellular mechanisms of toxicity. His most recent studies have concentrated on the use of in vitro systems to evaluate cell-specific injury, including the effects of low-level metal exposure on cell signaling and gene expression and the development of in vitro models to reflect in vivo toxicity. Tissues of interest are the liver, kidney, bladder, and prostate. Dr. Gandolfi earned a PhD in biochemistry and biophysics at Oregon State University.

David Kriebel is a professor in the Department of Work Environment of the University of Massachusetts Lowell. He is also codirector of the Lowell Center for Sustainable Production. Dr. Kriebel's research focuses on the epidemiology of occupational injuries, cancer, and nonmalignant respiratory disease. He has published on various aspects of epidemiologic methods, particularly on the use of quantitative exposure data in epidemiology. He has been active in developing dosimetric models to improve understanding of the effects of aerosols on the lungs. He teaches introductory and advanced courses in epidemiology, risk assessment, and research synthesis. Dr. Kriebel has served on several National Academies committees, including the Committee on Beryllium Alloy Exposures and the Committee to Review the Health Effects in Vietnam Veterans of Exposure to Herbicides. Dr. Kriebel received an ScD in epidemiology from the Harvard School of Public Health.

John B. Morris is the Board of Trustees Distinguished Professor, a professor of pharmacology and toxicology in the Department of Pharmaceutical Sciences, and interim dean of the University of Connecticut School of Pharmacy. His research focuses on toxicity of inhaled irritant vapors, irritants and asthma, regional uptake and metabolism of inspired vapors, physiologically based pharmacokinetic modeling, and risk assessment. Dr. Morris has served on the editorial boards of *Toxicological Sciences* and *Inhalation Toxicology* and on advisory panels for the National Institutes of Health, the Environmental Protection Agency, and the Department of Energy. He has also served as a member of the National Research Council Committee on Emergency and Continuous Exposure Guidance Levels for Selected Submarine Contaminants. Dr. Morris earned a PhD in toxicology from the University of Rochester.

Kent E. Pinkerton is a professor in the Department of Pediatrics in the School of Medicine and the Department of Anatomy, Physiology, and Cell Biology in the School of Veterinary Medicine of the University of California, Davis. He also serves as director of the university's Center for Health and the Environment. His research interests focus on the health effects of environmental air pollutants on lung structure and function, the interaction of gases and airborne particles in specific sites and cell populations of the lungs in acute and chronic lung injury, and the effects of environmental tobacco smoke on lung growth and development. Dr. Pinkerton has served as a member of the National Research Council Committee on Estimating Mortality Risk Reduction Benefits from Decreasing Tropospheric Ozone Exposure and the Committee for Review of the Army's Enhanced Particulate Matter Surveillance Project Report. Dr. Pinkerton received a PhD in pathology from Duke University.

Ivan Rusyn is a professor in the Department of Environmental Sciences and Engineering in the School of Public Health of the University of North Carolina (UNC) at Chapel Hill. He directs the Laboratory of Environmental Genomics and the Carolina Center for Computational Toxicology in the Gillings School of Global Public Health at UNC-Chapel Hill. He is a member of the Lineberger Comprehensive Cancer Center, the Center for Environmental Health and Susceptibility, the Bowles Center for Alcohol Studies, and the Carolina Center for Genome Sciences. Dr. Rusyn's laboratory focuses on the mechanisms of action of environmental toxicants, the genetic determinants of susceptibility to toxicant-induced injury, and computational toxicology. He has served on several National Research Council committees and is currently a member of the Committee on Use of Emerging Science for Environmental Health Decisions and the Committee on Toxicology. Dr. Rusyn received his MD from Ukrainian State Medical University in Kiev and his PhD in toxicology from UNC-Chapel Hill.

Toshihiro Shioda is an associate professor of medicine at Harvard Medical School and director of the Molecular Profiling Laboratory of Massachusetts General Hospital Cancer Center. Dr. Shioda's research focuses on how toxic substances in the environment affect the regulation of gene function in mammalian cells, including stem cells and germ cells. That work has included the use of cutting-edge technologies of genome analysis to determine how environmental endocrine disruptors and other toxicants act at the molecular level to disrupt genetic and epigenetic programming in human and mouse cells in live bodies and cell cultures. His laboratory is also exploring the molecular mechanisms of breast-cancer resistance to antiestrogen therapies. Dr. Shioda received his MD in Japan and earned a PhD in biochemistry from Hiroshima University Graduate School of Medicine.

Thomas J. Smith is a professor emeritus at the Harvard School of Public Health. His research interests are in the characterization of environmental and occupational exposures for studies of health effects and the investigation of the relationship between environmental exposure and internal dose. He has developed a toxicokinetic modeling approach for designing exposure evaluations for epidemiologic studies and is using the approach in a cohort study of lung-cancer mortality in the US trucking industry, in which workers are exposed to diesel exhaust. Dr. Smith is also involved in an exposure study of human metabolism of 1,3-butadiene. He has served as a member of the National Research Council Committee on Human Health Risks of Trichloroethylene and the Panel on Monitoring and the Institute of Medicine Committee on the Assessment of Wartime Exposure to Herbicides in Vietnam. Dr. Smith received a PhD in chemistry and environmental health from the University of Minnesota.

Meir Wetzler is chief of the Division of Leukemia of the Department of Medicine of Roswell Park Cancer Institute and a professor of medicine in the School of Medicine and Biomedical Sciences of the University at Buffalo, the State University of New York. Dr. Wetzler's research interests focus on the role of signal transducer and activation of transcription in leukemogenesis, the cellular and humoral immune response to leukemia-associated antigens, and cytogenetics in acute myeloid leukemia and acute lymphoblastic leukemia. Dr. Wetzler earned an MD from the Hebrew University Hadassah Medical School, Jerusalem, Israel.

Lauren Zeise is deputy director for scientific affairs in the California Environmental Protection Agency's Office of Environmental Health Hazard Assessment. Dr. Zeise oversees the department's scientific activities, which include the development of risk assessments, hazard evaluations, toxicity reviews, cumulative impacts analyses, frameworks and methods for assessing toxicity and cumulative impact, and activities in the California Environmental Contaminant Biomonitoring Program. Dr. Zeise was the 2008 recipient of the Society of Risk Analysis's Outstanding Practitioners Award. She has served on advisory boards and committees of the Environmental Protection Agency, the Office of Technology Assessment, the World Health Organization, and the National Institute of Environmental Health Sciences. Dr. Zeise has served on numerous National Research Council and Institute of Medicine committees. She is currently a member of the Committee to Review EPA's Draft Paper, State of the Science on Nonmonotonic Dose Response. Dr. Zeise received a PhD in environmental sciences from Harvard University.

Patrick Zweidler-McKay is section chief for pediatric leukemia and lymphoma and an associate professor in the Division of Pediatrics of The University of Texas M D Anderson Cancer Center. His interests are in developing targeted therapies for children who have leukemia and neuroblastoma, and his research laboratory is directed at understanding the critical pathways that contribute to leukemia and neuroblastoma. Clinically, he specializes in treating children who have particularly difficult or relapsed forms of leukemia and lymphoma, such as infant acute lymphoblastic leukemia and T-cell leukemia–lymphoma. He served as a member of the National Research Committee to Review the Draft IRIS Assessment on Formaldehyde. Dr. Zweidler-McKay earned a PhD in molecular biology and genetics and an MD from Temple University.

Appendix B

Statement of Task of the Committee to Review the Formaldehyde Assessment in the National Toxicology Program 12th Report on Carcinogens

A committee of the National Research Council will conduct a scientific peer review of the formaldehyde assessment presented in the National Toxicology Program (NTP) 12th Report on Carcinogens (RoC). The committee will identify and evaluate relevant, publicly available, peer-reviewed literature, with particular emphasis on literature published as of June 10, 2011, the release date of the 12th RoC. The committee will document its decisions for inclusion or exclusion of literature from its evaluation and will identify the set of information deemed most critical to the evaluation. The committee will apply independently the NTP's established RoC listing criteria to the scientific evidence from studies in humans, experimental animals, and other studies relevant to mechanisms of carcinogenesis and make independent level-of-evidence determinations with respect to the human and animal studies. The committee will integrate the level-of-evidence conclusions, and considering all relevant information in accordance with the RoC listing criteria, make an independent listing recommendation for formaldehyde and provide scientific justification for its recommendation.

Note: The NRC has an agreement with the Department of Health and Human Services to undertake a scientific peer review of the determinations concerning formaldehyde and styrene in the National Toxicology Program's 12th Report on Carcinogens (RoC). The expert committees appointed by the Academy for this assignment will follow standard Academy practices in carrying out their independent scientific reviews, which may include consideration of any and all issues that the committees and the Academy decide are necessary to carry out credible, independent, scientific evaluations of the two determinations, potentially including the criteria for the determinations. The statements of task for these two peer reviews were recently modified to make it clear that the NRC's assignment does not also include a separate review of the National Toxicology Program's listing criteria.

Appendix C

Exposure Assessment in Epidemiologic Carcinogenicity Studies

The purpose of this appendix is to describe the characteristics and attributes that the committee used to evaluate the exposure-assessment component of epidemiologic studies. The committee first provides introductory information on exposure assessment and its use to understand disease risk in defined populations. The committee then discusses components of an exposure assessment, including defining job titles; measuring exposures using time-weighted averages (TWAs), cumulative exposures, and peak exposures; strategies for exposure sampling; choosing an appropriate summary measure of exposure; and creating a job–exposure matrices (JEM) for use in cohort studies. The committee also discusses differences between exposure assessments for cohort studies of specific industries compared with general-population case–control studies and case–control studies nested within cohorts. The appendix ends with a summary table of the criteria that the committee used to assess the epidemiologic studies cited in Chapters 2 and 3.

INTRODUCTION TO EXPOSURE ASSESSMENT

The fundamental logic of epidemiologic analysis is the 2 × 2 table, in which one axis is the subjects' disease status (yes–no) and the other is their personal exposures (yes–no). The quality of a study and the strength of its conclusions depend strongly on exposure evaluation, in addition to its epidemiologic aspects. The basic goal of an exposure assessment is to evaluate the qualitative and quantitative discrimination of a study's exposure assignments. Different methods have different powers of discrimination.

By definition, exposure is personal and external to the individual. The points of entry for chemical exposure are the nose and mouth for inhalation, the mouth for ingestion, and the skin for dermal absorption. All three have the following dimensions: composition, intensity, and time course. Complexity arises along all three dimensions. First, it is rare that a person is exposed to only a single substance, such as formaldehyde; mixed exposures almost always occur. Some components of mixtures, such as formaldehyde vapors and paraformaldehyde particles emitted during embalming, may produce similar effects or may modify the effects of other substances, which may serve to confound relationships with disease. The physical form of formaldehyde vapor or particulate paraformaldehyde will strongly affect where it is deposited in the respiratory tract. Second, environmental concentrations are generally not constant in time or location. Sources of airborne formaldehyde are not continuous and steady. As a result, exposure varies in time and location. In addition, concentrations of individual mixture components may vary independently or correlate, depending on their sources. Third, the area near local emission sources, such as embalming fluid in body cavities, produces the highest, variable air concentrations that usually have considerable random temporal variation. These variations are the result of incomplete dilution and mixing processes in the breathing zone air over short periods, which produce approximately lognormal distributions for variations in consecutive concentration measurements. Regional concentrations away from local sources will be lower and are relatively more stable. Outdoor exposures commonly show hourly, daily, seasonal, or annual trends that are associated with weather, climate, source output, exposed subjects' activities, ventilation, and other factors. Those aspects of exposure are discussed in detail by Lippmann et al. (2003) and Smith and Kriebel (2010).

Epidemiologic researchers seek to exploit natural experiments in which large differences in environmental or occupational exposures occur among large groups of otherwise similar people. The exposure-assessment goal is to identify personal, occupational, or environmental factors that determine differences in exposure to the substance of interest (Checkoway et al. 2004). The gradients in exposure, if sufficiently large, can be used to determine whether there are corresponding gradients in disease risk that might be causally related. Useful occupational-exposure gradients can be produced by the nature of the subjects' jobs, tasks, or activities in the workplace and by the characteristics of work locations and materials used, such as formaldehyde solutions or paraformaldehyde powder. Similarly, characteristics of subjects' residences, commuting activities, food sources, and other determinants of environmental exposure can be used to define exposure groups for comparisons of risk. The rationale and quality of data used to assign exposure are important in determining the quality and reliability of the assignments. Blair and Stewart (1992) showed that improved quantitation of formaldehyde exposure tended to increase exposure gradients and sharpen estimates of relative risk. Exposure assignments that are imprecise can result in individuals being categorized into the wrong category of the exposure gradient and the epidemiologic study analysis table. Misclassification reduces the apparent relative risk and may produce misleading conclusions.

An important step in the use of exposure data for an epidemiologic study is the construction of a summary measure of exposure (Smith and Kriebel 2010). When semiquantitative or quantitative data on intensity and duration of exposure for study participants are available, these must be summarized—usually in a single number—to be used in an epidemiologic model to assess the strength of the exposure–risk association. The choice of which summary measure[1] to use should ideally be based on biologic hypotheses about the underlying causal mechanism. In practice, this information is often lacking, so indices are tested and goodness-of-fit data are used to assess which metric is more likely to be (approximately) correct. Unfortunately, the precision of exposure metrics is often low. As a result, it is not possible to determine if one metric is substantially better than another. Such a distinction would be highly useful.

COMPONENTS OF AN EXPOSURE ASSESSMENT

Industrial hygienists are trained to recognize hazards in an occupational setting, how to evaluate those hazards, and how to reduce or control exposures. Part of their expertise includes analyzing workplace organization and defining jobs and their job titles, work activities, or work locations in specific industries. Typically, job titles, department titles, and work locations will be collected from an individual's job history, which is usually held in company records. Company records also commonly contain extensive data on the site of the industrial operations, including plant maps, locations of major equipment and operations where exposures would have taken place, the raw materials used and products and byproducts produced, and the emission-control equipment used and when it was installed. Industrial hygienists are also trained to take measurements of chemical exposures of individual workers and to assess the quality of available measurement data. Industrial hygienists who are interested in epidemiologic research may also obtain training in the estimation of historic exposures suitable for the extrapolation of long-term past exposures associated with chronic disease. Table C-1 shows how basic knowledge about sources of formaldehyde emissions, the physical setting, the type of job, the job location, and the activities that make up the job can be used to make useful distinctions that discriminate among different levels of exposure. Various approaches have been used to define differences between scenarios of high, medium, low, and no exposure. The various approaches are not equally useful for discriminating exposures with minimum misclassification. High-quality exposure assessments can accomplish that by using the strategies outlined in the section below.

Job Titles

Job titles are labels used by management for personnel functions to organize work activities. In some cases those work activities may have close links with exposure, but the job titles may or may not be associated with exposures depending on how the work activities were distributed across the job titles. The

---

[1]Typical summary measures of exposure include average exposure, duration of exposure, cumulative exposure, and various measures of peak exposure.

TABLE C-1 Distinctions between Different Levels of Exposure

High exposure	• Job histories that include job titles, tasks, or activities that take place close to sources of concentrated formaldehyde emissions can provide information on the potential for high exposures. • Job-site data can provide information on work areas, equipment, and chemicals that are heavily used and handled often. • Emission and work-area measurement data indicate general high-exposure levels, and poorly mixed, concentrated emissions may produce substantial peak[a] exposures. • Absence of emission controls or poor ventilation[b] in a setting in which vapor can accumulate, such as a warehouse where materials off-gas incompletely or where reactive chemical coatings are present, can lead to high mean concentrations but less extreme peaks.
Medium exposure	• Job and work-area data identify tasks or activities that take place at a distance from sources of concentrated formaldehyde emissions. These exposures are difficult to define qualitatively or semiquantitatively, and data is often absent. The central tertile of a measured exposure distribution is prone to misclassification into both the high-exposure and low-exposure groups. • Often insufficient job or work-area data or unevaluated assumptions lead to misclassification of exposures as high or low.
Low exposure	• Jobs, tasks, or activities with only brief periods when formaldehyde vapors are present, and the work location is distant from the sources. • Physical separation of work areas from areas with emission sources. • Good ventilation prevents vapors from accumulating in the area.
Exposure controls	• If respirators or ventilation engineering systems[c] are used, it is important to find documentation in plant records that describe when the controls started to be used and how effective they were at reducing or eliminating exposure. • Respirators or ventilation systems were usually effective after the middle 1970s. Before then, they were less effective.
No exposure	• Work in a setting that has no sources of formaldehyde emissions.

[a]Peak exposures are short-duration (approximately 15 minutes, but the precise length is often not defined), high-concentration (for example, >2–4ppm) exposures. They may be defined by the limitations of measurement methods.
[b]Ventilation is the amount of air flowing through a work space from windows and doors and by forced ventilation.
[c]Ventilation engineering systems, including fans, ductwork, hoods, and enclosures that provide ventilation, control and minimize airborne emissions.

work histories of study subjects can be a useful link with occupational exposure conditions, but that link and the exposure conditions must be defined with an exposure assessment. A given job title may be associated with substantially different work activities and exposures in different companies or during different historical periods. For example, a chauffeur today may drive a limousine, but a chauffeur before the 1960s was often a truck driver, and that required a chauffer's license. Therefore, a person with a chauffeur's license in the 1960s may have had very different exposures compared to a chauffeur today. Industrial hygiene expertise and data from long-term workers is required to translate job and work location information into exposure assignments.

A widely used set of standardized job descriptions—the International Standard Industrial Classification—has been developed by the United Nations (UN 2008). A similar set of more specifically defined occupations—the Dictionary of Occupational Titles—has been developed by the US Department of Labor (DOL 1991). Because the UN and Department of Labor job titles are broad, their link to exposures is often weak, and misclassification is common. A simple and specific job title may be satisfactory for exposure classification if it unequivocally links with an exposure situation. For example, a person whose job title is "embalmer" often uses solutions with high concentrations of formaldehyde while embalming bodies in a small, poorly ventilated room. Those conditions will consistently lead to exposure to high concentrations of formaldehyde vapor. Other, more generic and broad titles, such as "mortician" or "funeral director", also may involve embalming but less often, and embalming is not one of the main job activities. Thus, the title "mortician" is broader and includes more people but leads to more misclassification and much less discrimination for formaldehyde exposure than the title "embalmer".

Epidemiologists and exposure assessors have addressed the poor specificity of standard job titles by adding sets of titles that are specific for the industries under study. They have also added questions to questionnaires and interview guides to ask about specific jobs, activities, and substances that are expected to be present, such as "embalmer", "embalming", and "formaldehyde and paraformaldehyde". They may also ask about irritation and odors that distinguish particular substances. The utility of such questions depends on the subjects' knowing the names and other properties of an agent. It is common for workers not to know the names of substances to which they are exposed; for example, they may know only that they use a clear liquid in a blue can to clean up grease and oil. The identity of the liquid must come from other sources, such as material safety data sheets kept by the company.

Professional requirements, unionization, and certifications can improve the exposure specificity of job definitions. Legal requirements for embalming and preparation of bodies for interment reduce the variation in exposure opportunities. Lower-level nonprofessional jobs—such as laborer, technician, and assistant—often have poorly defined tasks and work locations and are difficult to classify with respect to exposure.

Exposure Measurements

Formal quantitative measurement is the best way to determine to what and where people are exposed. The accuracy and precision of exposure measurements have improved greatly, particularly since the 1970s when extensive exposure surveys and routine monitoring began (Stewart et al. 2000) and when standards were established for allowable exposures in the United States and other countries (Stewart et al. 1996; Symanski et al. 1998). Current methods for exposure measurements were developed by the National Institute for Occupational Safety and Health. The methods have been standardized to measure allowable exposures or emissions for regulatory purposes and they are used by the US Environmental Protection Agency for measuring formaldehyde vapor.[2] The numbers of samples collected have also increased because of concern about exposure variability. In many cases, few or no historical exposure data have been available for long-term health studies. However, increasingly sophisticated extrapolation strategies have been developed (discussed below).

Time-Weighted Average

Inhaling a time-varying concentration at a fixed rate, such as 10 L/min (light exercise) for a specific time period (such as an 8-hour work day) produces a TWA concentration over the period of exposure. Similarly, drawing air or water into a collector at a fixed flow rate (volume per unit time) for a defined period produces a TWA sample in the collector because an equal volume is passed through each minute (Δt) and each unit of volume contributes material in proportion to the concentration:

$$TWA = SUM(C[i]\Delta t)/SUM(\Delta t) = SUM(C[i])/N,$$

Where C is the concentration of the substance, i is the period, and N identifies the number of periods; $T = N\Delta t$. That is analogous to inhalation at a fixed breathing rate and is a good dose metric for exposed subjects during their work period (shift).

Cumulative Exposure

Cumulative exposure is perhaps the most commonly used summary measure of exposure in occupational epidemiology of chronic diseases. Cumulative exposure is defined as the product of the average exposure concentration (C)

---

[2]Publications on formaldehyde methods of both agencies can be obtained at http://www.ntis.gov/search/index.aspx.

multiplied by the duration of exposure (T). The theoretical basis for the widespread use of cumulative exposure is Haber's rule (also called Haber's law), which posits that within an appropriate range for inhaled toxicants, all combinations of C and T with the same value will all produce the same effect (Belkebir et al. 2011). The rule breaks down outside narrow ranges of C and T. The rule implies that high exposures for short durations produce the same effects as low exposures for long durations, which may not be true. This rule also implicitly assumes that all toxic processes have no thresholds or lags for responses.

Some of the formaldehyde cancer studies reviewed by the committee used cumulative exposure to summarize occupational exposures across each study subject's entire work history. If Haber's rule holds for the carcinogenic effects of formaldehyde, then summarizing exposure histories using cumulative exposure will not introduce any exposure misclassification. However, if a few years of high exposure early in a subject's work life are more important for cancer risk than many years of low exposure, then using cumulative exposure will introduce misclassification and reduce the likelihood of detecting an association.

In studies of occupational exposure in which the intensity of exposure has not been measured, duration of work is sometimes used as a surrogate for cumulative exposure. Duration of exposure will be proportional to cumulative exposure when the average exposure is approximately the same for all members of the cohort, so that the only person-to-person variability in cumulative exposure derives from differences in duration of exposure. Because this assumption is not likely to be true, there can be substantial misclassification within and between exposure groups in their average durations of work and exposure, which will probably bias the results toward the null (Kriebel et al. 2007). On the other hand, differences in the intensity of exposure among groups in early studies, such as those between embalmers and other funeral workers, were probably quite large so that substantial differences in risk by years of work or exposure would be expected.

Peak Exposure

High-intensity but short-duration exposures are called peaks. Peaks are of interest because they are much higher concentrations than the mean exposure and as a result may exceed a minimum intensity needed to cause an acute effect that has threshold or nonlinear pharmacodynamics. Peaks are quantified by the product of concentration at the point of entry and duration ($C \times \Delta t$), where Δt is a short period, such as 15 minutes. The smallest $C \times \Delta t$ for a biologically relevant peak is implicitly the acute dose needed to produce a minimum effect. The range of definitions of a peak used by studies can be broad because the minimum effective dose is not known. The concentration of formaldehyde reported to cause upper airway irritation, 2–4 ppm, has commonly been used to define the minimum peak concentration, but it is not known whether this is relevant for carcinogenesis. As a practical approach, the limitations of the industrial-hygiene measurement techniques have often been used to define the minimum intensity and Δt of peaks. Allowable peak exposures set by regulatory agencies are based on characteristics and limitations of monitoring methods. For example, formaldehyde concentrations that exceed 2 ppm for 15 minutes exceed the Occupational Safety and Health Administration short-term exposure limit (OSHA 2014). Historically, the 15-minute duration was chosen because 15 minutes of sampling were needed to collect enough material for acceptable measurement precision. Biologically important peaks of shorter duration might produce upper respiratory tissue effects. The acute-dose definition, C × Δt, also breaks down at the extremes of concentration and short duration because of pharmacokinetic and physiologic limits on uptake, transport, activation and deactivation, and removal.

Some jobs or activities have clear opportunities for peak exposures; for example, embalmers work with high-concentration sources nearby, but others do not, such as workers in a garment warehouse. In a garment warehouse, the incomplete polymerization of a fabric's permanent-press treatment is the source of formaldehyde. Emissions from a single garment are limited, but there are many hundreds or thousands of garments throughout the warehouse. Thus, concentrations do not vary widely, and there are not expected to be high peaks, but average concentrations can be high; high peaks and high-TWA exposures do not necessarily occur together. An exposure assessment specifically designed for the task is needed to determine where peaks may occur.

Assessment of peak exposures for jobs and work activities requires considerable detailed information. Only large, extensive studies have collected the necessary data and measurements to estimate the intensity, frequency, and duration of situations with peak exposures, such as the National Cancer Institute cohort studies of the US chemical industries and the funeral industry (Beane Freeman et al. 2009; Hauptmann et al. 2009). Peaks also contribute to TWA exposures, but they are of short duration and the correlation between peaks and cumulative exposures tends to be weak (Blair and Stewart 1992). Thus, if peaks are causally related to cancer risk, then using average exposure or cumulative exposure metrics will introduce misclassification. However, the peak-exposure metrics are of limited precision and may not be sufficient to distinguish a peak mode of action from a cumulative mode of action. As stated in Chapter 3, it is expected that, on average, choosing the wrong metric will result in an underestimation of an association if one exists (Checkoway et al. 2004).

Sampling Strategy

Exposures are generally highly variable in time and location, but it is impractical to measure them all continuously. Therefore, measurement of personal and location exposures use several types of statistical sampling strategies. Sampling strategies have changed considerably with the development of personal TWA sampling (a small pump and lapel collector) and the implementation of the

Occupational Safety and Health Act of 1970. Therefore, historical sampling data from before the 1970s need to be carefully evaluated and sometimes adjusted (Corn 1992).

The most useful strategy for epidemiology is the collection of random personal samples from different exposure groups. They should be collected at or near the route of entry, such as in the breathing zone, and for the whole duration of exposure. Fixed-location ("area") samples or stationary samples have been widely collected in places where people may be present at some times, but these may lead to overestimation of exposure if they are taken closer to sources than where people are normally located. They may lead to underestimation of exposures if people are present for only short periods relative to the duration of the sampling or if people normally are closer to sources compared with the location of the monitors. If area samplers are used consistently with the same strategy, they tend to produce samples that are proportional to personal exposures, and the proportionality can be estimated on the basis of the ratio of concurrent personal to area sampling.

Job–Exposure Matrix for a Cohort Study in a Single Company

JEM methods were developed by several investigators, including Stewart et al. (1996). The approach used by industrial hygienists to develop JEM assignments for cohort studies is summarized below.

1. Job titles and plant or worksites associated with jobs are abstracted from company work histories, or cases and controls or their proxies are interviewed.

2. Jobs, worksites, processes, and work rooms are located on plant diagrams, and historical changes are also recorded.

3. Industrial hygienists with knowledge of the industry visit plants for walk-throughs and discuss operations, processes, materials, jobs, and historical changes with long-term workers and supervisors. This information is used to develop a plant history.

4. Industrial hygienists collect all available exposure measurements, personal data, and area data. The amount and quality of data will vary widely by date, plant area, and job. The data also may be limited by plant closures and loss of records.

5. If possible, industrial hygienists conduct field studies to measure exposures and conduct studies of job activities and task exposures, as was done by Stewart et al. (1992) for embalmers.

6. In some cases, sufficient data are available to develop detailed statistical models that can be used to estimate exposures. An example is the work by Hornung et al. (1996). Alternatively, extrapolation models have been developed on the basis of physical principles and extrapolations from current conditions

backward in time (Stewart et al. 1996; Tielemans et al. 2008; Fransman et al. 2011).

7. The exposure information and other data that are collected are used to develop JEM tables for each unique job title and work location by year. Estimates are made of the TWA and of the potential for peak exposures, the frequency of such exposures, and the intensity for each substance of interest. A good example is Blair et al. (1986). Some JEMs may be less complete than others, and this will limit the types of exposure estimates that are possible and may increase the amount of misclassification and thus reduce the ability of a study to detect small risks. The plant-history documentation and exposure estimates are sent to participating plants for technical review by company engineers and industrial hygienists to verify their accuracy.

EXPOSURE ASSESSMENT FOR CASE-CONTROL STUDIES

Exposure assessment for case–control studies that draw their subjects from the general population is difficult because they generally rely on recalled job titles and industries. Even when recall is accurate, there will be a loss of information because the occupation and industry information must be coded using a broad classification system such as the International Standard Classification of Occupations (ISCO) and the International Standard Industrial Classification. An example is a worker reporting he was a salesman for automotive parts. His position might be coded using ISCO code 43 for "male technical salesmen, commercial travelers, and manufacturer's agents." That broad grouping will usually have little specificity for a particular chemical exposure of interest, such as formaldehyde. In addition, the distribution of occupations and exposures depends heavily on the distribution of local industries and the prevalence of formaldehyde users in a region. That problem can be reduced by choosing a base population that has a large prevalence of an industry of interest. The study by Luce et al. (2002) drew from areas that had large industries processing wood, which resulted in few subjects who were exposed to formaldehyde without also being exposed to wood dust. Some investigators, such as Luce et al. (2002), improve their specificity by preparing an additional detailed questionnaire on formaldehyde-related jobs. However, as noted earlier, workers or their next of kin often do not know their exposures to specific chemicals with which they worked.

Where there are no exposure data for the study sites, expert or professional industrial-hygiene judgment is often used to estimate who has been exposed and their degree of exposure. Jobs, work activities, and work areas need to be evaluated to achieve specificity. Questionnaire data collected from the subjects, their peers, or next of kin are often evaluated by industrial hygienists familiar with local conditions to assess job or area exposures. There have been a number of evaluations of such expert judgment. For example, Luce et al. (1993) conducted an evaluation of expert judgment used in their population-based case–control study of sinonasal cancer.

Appendix C 195

Formaldehyde's irritant properties are readily recognized, which may make identifying the presence of this specific exposure easier. Coggon et al. (1984) used the presence of substantial irritation as a marker of "high" exposure in areas where formaldehyde was known to be used. Unfortunately, this approach is limited by the broad variation in human sensitivity to irritants and by the tendency for people to acclimatize after a period of low to moderate exposure. Also, sensitive individuals may leave the workplace while long-term workers may be self-selected for being relatively insensitive to the irritant effects. As a result, worker appraisals of irritation may underestimate the exposures.

Case–control studies that are drawn from members of an exposed cohort (that is, "nested" case–control studies) have an advantage for exposure assessment because exposures in the source cohort may already have been assessed, and detailed exposure assignments may be available (Checkoway et al. 2004). That can make a study very discriminating for specific agents and long periods.

INFORMATION USED TO EVALUATE EXPOSURE ASSESSMENTS

The committee evaluated five aspects of each epidemiologic study reviewed in Chapters 2 and 3 to determine the quality of discrimination and the utility of an exposure assessment. Those aspects are the expertise of the investigators, the assessment type (such as, personal monitoring or JEM methods), the availability of key data (including job history, site information, and sampling measurements), the potential for misclassification (both qualitative and quantitative), and, where possible, the evaluation of the peak exposures. High quality in the first four aspects of an assessment produces a strong exposure assessment with high discrimination for long-term exposures. Table C-2 shows the information the committee used to review and evaluate the epidemiologic studies cited in Chapters 2 and 3.

TABLE C-2 Information Used to Evaluate Exposure Assessment Components of Epidemiologic Studies in Chapters 2 and 3

Overall Method	Exposure-Assessment Components					Exposure Assignments	Discrimination of Exposure Differences Between Categories
	Job-History Data	Site Data and Industrial-Hygiene Evaluation	Sampling Data	Extrapolation of Past Exposures		
Qualitative—broad occupational groups and industries in a region	None	None	None	None	Yes—qualitative	**Low**—few exposed in broad job groups; strong tendency to overestimate number exposed; likely large misclassification
Semiquantitative—specific jobs in one industry	Yes—job descriptions, interviews, questionnaires, and proxies; industrial hygienist uses professional judgment to assess exposures	None—many worksites or no data on specific sites	None or very limited for the industry	None or maybe some data on time trends; industrial hygienist uses professional judgment to assess past exposures	Yes—Semiquantitative in years of exposure	**Moderate**—specific job titles and work site data; limited measurements; likely much overlap between categories **High**—specific jobs with defined exposures and limited overlap of low and high categories
Quantitative—specific jobs or areas in a company	Yes—detailed company records	Yes—extensive data on operations, sites, and job activities	Yes—extensive for high-exposure jobs or areas over time	Yes—detailed strategies and modeling; industrial hygienist uses professional judgment to assess past exposures	Quantitative by substance, job or area, and period according to dose metrics	**Moderate**—if specific job, area, or sampling data are limited; likely overlap between groups **High**—limited overlap between low- and high-exposure categories

REFERENCES

Beane Freeman, L.E., A. Blair, J.H. Lubin, P.A. Stewart, R.B. Hayes, R.N. Hoover, and M. Hauptmann. 2009. Mortality from lymphohematopoietic malignancies among workers in formaldehyde industries: The National Cancer Institute Cohort. J. Natl. Cancer Inst. 101(10):751-761.

Belkebir, E., C. Rousselle, C. Duboudin, L. Bodin, and N. Bonvallot. 2011. Haber's rule duration adjustments should not be used systematically for risk assessment in public health decision-making. Toxicol. Lett. 204(2-3):148-155.

Blair, A., and P.A. Stewart. 1992. Do quantitative exposure assessments improve risk estimates in occupational studies of cancer. Am. J. Ind. Med. 21(1):53-63.

Blair, A., P. Stewart, M. O'Berg, W. Gaffey, J. Walrath, J. Ward, R. Bales, S. Kaplan, and D. Cubit. 1986. Mortality among industrial workers exposed to formaldehyde. J. Natl. Cancer Inst. 76(6):1071-1084.

Checkoway, H., N. Pearce, and D. Kribel. 2004. Research Methods in Occupational Epidemiology, 2nd Ed. Oxford: Oxford University Press.

Coggon, D., B. Pannett, and E.D. Acheson. 1984. Use of job-exposure matrix in an occupational analysis of lung and bladder cancers on the basis of death certificates. J. Natl. Cancer Inst. 72(1):61-65.

Corn, M. 1992. Historical perspective on approaches to estimation of inhalation risk by air sampling. Am. J. Ind. Med. 21(1):113-123.

DOL (U.S. Department of Labor). 1991. Dictionary of Occupational Titles, 4th Ed. [online]. Available: http://www.oalj.dol.gov/libdot.htm [accessed Jan. 31, 2014].

Fransman, W., M. Van Tongeren, J.W. Cherrie, M. Tischer, T. Schneider, J. Schinkel, H. Kromhout, N. Warren, H. Goede, and E. Tielemans. 2011. Advanced Reach Tool (ART): Development of the mechanistic model. Ann. Occup. Hyg. 55(9):957-979.

Hauptmann, M., P.A. Stewart, J.H. Lubin, L.E. Beane Freeman, R.W. Hornung, R.F. Herrick, R.N Hoover, J.F. Fraumeni Jr., A. Blair, and R.B. Hayes. 2009. Mortality from lymphohematopoietic malignancies and brain cancer among embalmers exposed to formaldehyde. J. Natl. Cancer Inst. 101(24):1696-1708.

Hornung, R.W., R.F. Herrick, P.A. Stewart, D.F. Utterback, C.E. Feigley, D.K. Wall, D.E. Douthit, and R.B. Hayes. 1996. An experimental design approach to retrospective exposure assessment. Am. Ind. Hyg. Assoc. J. 57(3):251-256.

Kriebel, D., H. Checkoway, and N. Pearce. 2007. Exposure and dose modeling in occupational epidemiology. Occup. Environ. Med. 64(7):492-498.

Lippmann, M., B.S. Cohen, and R.B. Schlesinger. 2003. Environmental Health Science: Recognition, Evaluation, and Control of Chemical and Physical Health Hazards. Oxford: Oxford University Press.

Luce, D., M. Gerin, F. Berrino, P. Pisani, and A. Leclerc. 1993. Souces of discrepancies between a job exposure matrix and a case expert assessment for occupational exposure to formaldehyde and wood-dust. Int. J. Epidemiol. 22(suppl.2):S113-120.

Luce, D., A. Leclerc, D. Begin, P.A. Demers, M. Gerin, E. Orlowski, M. Kogevinas, S. Belli, I. Bugel, U. Bolm-Audorff, L.A., Brinton, P. Comba, L. Hardell, R.B. Hayes, C. Magnani, E. Merler, S. Preston-Martin, T.L. Vaughan, W. Zheng, and P. Boffetta. 2002. Sinonasal cancer and occupational exposures: A pooled analysis of 12 case-control studies. Cancer Causes Control 13(2):147-157.

OSHA (Occupational Safety and Health Administration). 2014. Formaldehyde standards [online]: Available: https://www.osha.gov/pls/oshaweb/owadisp.show_document?p_id=10075&p_table=STANDARDS [accessed June 19, 2014].

Smith, T.J. and D. Kriebel. 2010. A Biologic Approach to Environmental Assessment and Epidemiology. New York: Oxford University Press.
Stewart, P.A., R.F. Herrick, C.E. Feigley, D.F. Utterback, R. Hornung, H. Mahar, R. Hayes, D.E. Douthit, and A. Blair. 1992. Study design for assessing exposures of embalmers for a case-control study. Part I. Monitoring results. Appl. Occup. Environ. Hyg. 7(8):532-540.
Stewart, P.A., P.S. Lees, and M. Francis. 1996. Quantification of historical exposures in occupational cohort studies. Scand. J. Work Environ. Health 22(6):405-414.
Stewart, P.A., R. Carel, C. Schairer, and A. Blair. 2000. Comparison of industrial hygienists' exposure evaluations for an epidemiologic study. Scand. J. Work Environ. Health 26(1):44-51.
Symanski, E., L.L. Kupper, I. Hertz-Picciotto, and S.M. Rappaport. 1998. Comprehensive evaluation of long-term trends in occupational exposure: Part 2. Predictive models for declining exposures. Occup. Environ. Med. 55(5):310-316.
Tielemans, E., T. Schneider, H. Goede, M. Tischer, N. Warren, H. Kromhout, M. Van Tongeren, J. Van Hemmen, and J.W. Cherrie. 2008. Conceptual model for assessment of inhalation exposure: Defining modifying factors. Ann. Occup. Hyg. 52(7):577-586.
UN (United Nations). 2008. International Standard Industrial Classification (ISIC) of All Economic Activities, Rev. 4, ST/ESA/STAT/SER.M/4/Rev.4. Department of Economic and Social Affairs, Statistics Division, UN, New York [online]. Available: http://unstats.un.org/unsd/publication/seriesM/seriesm_4rev4e.pdf [accessed Jan. 31, 2014].

Appendix D

Literature-Search Strategies Completed in Support of the Committee's Independent Assessment of Formaldehyde

The committee used the background document for formaldehyde as a starting point for its independent assessment of formaldehyde. In addition, it undertook several literature searches to identify any relevant literature that was published after the release of the 12th RoC. Each search covered the period from January 1, 2009 (the year in which the background document for formaldehyde was published; Bucher 2013), to November 8, 2013. Databases searched were PubMed, MEDLINE (Ovid), Embase (Ovid), Scopus, and Web of Science. The general topics of the searches include epidemiology, experimental-animal studies, and mechanisms of carcinogenicity (specifically, genotoxicity, mutagenicity, and hematologic effects). Each search was originally run on May 10, 2013, and updated on November 8, 2013. The search strategies, exclusion strategies, and number of resulting studies are described below.

CANCER STUDIES IN HUMANS

The committee established exclusion criteria and a literature-search strategy to identify studies in humans (Box D-1). The search resulted in 245 articles, as depicted in Figure D-1. National Research Council staff reviewed the titles and abstracts and excluded 221 as not relevant on the basis of the exclusion criteria. That left 24 articles that were identified as probably or possibly relevant. Two committee members reviewed the titles and abstracts and found 20 more that could be excluded. That left four articles that were considered as part of the committee's independent assessment.

EXPERIMENTAL-ANIMAL STUDIES

The literature search for publications of animal carcinogenicity bioassays yielded 280 results. The search terms are described in Box D-2, and a search tree

> **BOX D-1** Exclusion Criteria and Search Strategy for Human Studies
>
> Exclusion Criteria
>
> - The study did not evaluate ambient or occupational exposures of humans to formaldehyde.
> - The study did not evaluate health effects related to carcinogenesis or genetic damage.
> - The publication was already cited in the substance profile for formaldehyde in the National Toxicology Program 12th Report on Carcinogens.
> - The publication did not include primary data.
>
> Search Strategy
>
> *PubMed*: [("Formaldehyde"[Title/Abstract]) AND ("Neoplasms"[MeSH] OR neoplasms OR cancer OR carcinogenic or tumor) AND ("Epidemiology"[MeSH] OR "Epidemiologic Studies"[MeSH] OR epidemiolog* OR case-referent OR "Occupational Exposure"[MeSH] OR workers OR cohort)]. Search run on 05-10-2013 and updated on 11-08-2013; limited to 2009–2013.
>
> *Medline and Embase*:[(formaldehyde.ab. or formaldehyde.ti.) and (neoplasms/ or neoplasms.mp. or cancer.mp. or carcinogenic.mp. or tumor.mp.) and (epidemiology/ or epidemiologic studies/ or epidemiolog*.mp. or case-referent.mp. or occupational exposure or coworkers.mp. or cohort.mp.)]. Search run on 05-10-2013 and updated on 11-08-2013; limited to 2009–2013.
>
> *Scopus:* [("Formaldehyde") AND ("neoplasms" OR "cancer" OR "carcinogenic" OR "tumor") AND ("epidemiology" "epidemiologic studies" OR "epidemiolog*" OR "case-referent" OR "occupational exposure" OR "workers" OR "cohort")]. Search run on 05-10-2013 and updated on 11-08-2013; limited to 2009–2013.
>
> *Web of Science:*[("Formaldehyde") AND ("neoplasms" OR "cancer" OR "carcinogenic" OR "tumor") AND ("epidemiology" OR "epidemiologic studies" OR "epidemiolog*" OR "case-referent" OR "occupational exposure" OR "worker" OR "cohort")]. Search run on 05-10-2013 and updated on 11-08-2013; limited to 2009–2013.

representing the results is depicted in Figure D-2. A committee member and National Research Council staff independently screened the titles for potential papers reporting on animal cancer bioassays. No studies that exposed experimental animals to formaldehyde and evaluated them for the presence of tumors were identified. Thus, the committee's independent evaluation of the evidence of formaldehyde carcinogenicity in experimental animals relies on studies that were available to the National Toxicology Program when it conducted its review in 2011.

Appendix D 201

STUDIES OF MECHANISM OF CARCINOGENESIS

Genotoxicity and Mutagenicity

It is generally accepted that formaldehyde, because of its high reactivity, is genotoxic and may cause mutations and other cytogenetic effects that are collectively recognized as a mutagenic mode of action. Multiple types of DNA damage and later heritable changes in the cellular genome have been identified as possible consequences of exposure of DNA, cells, or tissues in vivo to formaldehyde. Thus, the literature-search terms pertinent to this mode of action were defined broadly to represent a variety of end points (Box D-3). The search was informed by a recently published case study of applying the principles of the systematic review to identify and present mechanistic evidence in human health assessments (Kushman et al. 2013).

The literature search for this topic resulted in 554 publications. The literature tree in Figure D-3 shows how the initial search results were narrowed down to 83 publications by National Research Council staff using publication titles and abstracts. The remaining publications were evaluated by two committee members using the titles, abstracts, and full text. In the end, 54 studies were considered relevant to the committee's independent assessment.

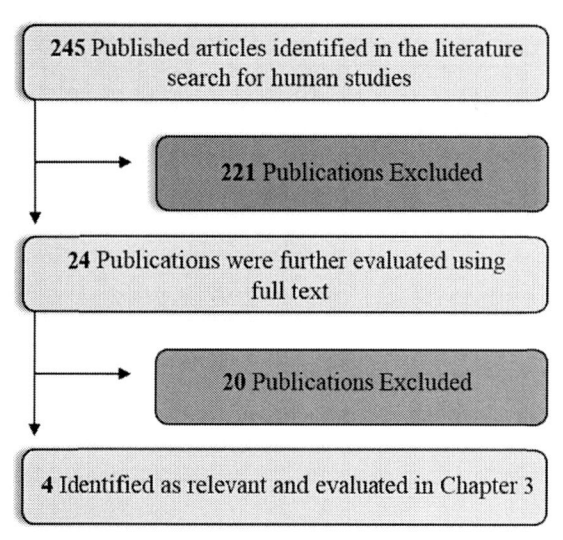

FIGURE D-1 Literature tree for human studies search. See Box D-1 for a description of the exclusion criteria and search strategy.

BOX D-2 Exclusion Criteria and Search Strategy for
Experimental-Animal Studies

Exclusion Criteria

- The study did not evaluate formaldehyde exposures in animal models.
- The study did not evaluate the incidence of tumors.
- The publication was already cited in the substance profile for formaldehyde in the National Toxicology Program 12th Report on Carcinogens.
- The publication did not include primary data.

Search Strategy

Pubmed: [("Formaldehyde"[Title/Abstract]) AND ("Neoplasms"[MeSH] OR "Carcinogen"[MeSH] OR cancer OR Foci OR Malignant* OR Oncogenic* OR Tumor OR Tumorigenic*) AND ("Animals"[MeSH] OR mice OR rats)]. Search run on 05-10-2013 and updated on 11-08-2013; limited to 2009–2013.

Medline and Embase: [(formaldehyde.ab. or formaldehyde.ti.) AND (neoplasms/ or carcinogens/ or cancer.mp. or foci.mp. or malignan*.mp. or oncongenic.mp. or tumor.mp. or tumorgenic*.mp.) AND (animals/ or mice.mp. or rats.mp.)]. Search run on 05-10-2013 and updated on 11-08-2013; limited to 2009–2013.

Scopus: [("Formaldehyde") AND ("neoplasms" OR "carcinogens" OR "cancer" OR "foci" OR "malignan*" OR "oncogenic*" OR "tumor" OR "tumorigenic*") AND ("animals" OR "mice" OR "rats")]. Search run on 05-10-2013 and updated on 11-08-2013; limited to 2009–2013.

Web of Science: [("Formaldehyde") AND ("neoplasms" OR "carcinogens" OR "cancer" OR "foci" OR "malignan*" OR "oncogenic*" OR "tumor" OR "tumorigenic*") AND ("animals" OR "mice" OR "rats")]. Search run on 05-10-2013 and updated on 11-08-2013; limited to 2009–2013.

---

280 Published articles identified in the literature search for experimental animal studies

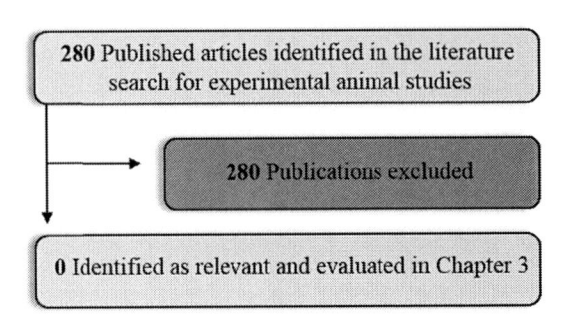

→ 280 Publications excluded

0 Identified as relevant and evaluated in Chapter 3

FIGURE D-2 Literature tree for experimental-animal studies search. See Box D-2 for a description of the exclusion criteria and search strategy.

Appendix D

> **BOX D-3** Exclusion Criteria and Search Strategy for Genotoxicity and Mutagenicity Mechanisms of Carcinogenesis
>
> Exclusion Criteria
>
> - The study did not evaluate health effects of formaldehyde or its metabolites known to be formed in humans.
> - The study evaluated cellular, biochemical, or molecular effects not relevant to the carcinogenesis or the mechanistic event under consideration.
> - The publication did not contain primary data.
> - The study did not include information sufficient to determine what species were studied or what experimental methods were used.
>
> Search Strategy
>
> *PubMed:* [("Formaldehyde"[Title/Abstract]) AND ("Mutation"[MeSH] OR "Cell Transformation, Neoplastic"[MeSH] OR "Cytogenetic Analysis"[MeSH] OR "Mutagens"[MeSH] OR "Oncogenes"[MeSH] OR "Genetic Processes"[MeSH] OR chromosom* OR clastogen* OR "genetic toxicology" OR "strand break" OR "unscheduled DNA synthesis" OR "DNA damage" OR "DNA adducts")]. Search run on 05-10-2013 and updated on 11-08-2013; limited to 2009–2013.
>
> *Medline and Embase:* [(formaldehyde.ab. or formaldehyde.ti.) and (mutation/ or cell transformation/ or cytogenetic analysis/ or mutagens/ or oncogenes/ or genetic processes or chromosom*.mp. or clastogen*.mp. or genetic toxicology.mp. or strand break.mp. or unscheduled DNA synthesis.mp. or DNA damage.mp. or DNA adducts.mp.)]. Search run on 05-10-2013 and updated on 11-08-2013; limited to 2009–2013.
>
> *Scopus:* [("Formaldehyde") AND ("mutation" OR "cell transformation, neoplastic" OR "cytogenetic analysis" OR "mutagens" OR "oncogenes" OR "genetic processes" OR "chromosom*" OR "clastogen*" OR "genetic toxicology" OR "strand break" OR "unscheduled DNA synthesis" OR "DNA damage" OR "DNA adducts")]. Search run on 05-10-2013 and updated on 11-08-2013; limited to 2009–2013.
>
> *Web of Science:* [("Formaldehyde") AND ("mutation" OR "cell transformation, neoplastic" OR "cytogenetic analysis" OR "mutagens" OR "oncogenes" OR "genetic processes" OR "chromosom*" OR "clastogen*" OR "genetic toxicology" OR "strand break" OR "unscheduled DNA synthesis" OR "DNA damage" OR "DNA adducts")]. Search run on 05-10-2013 and updated on 11-08-2013; limited to 2009–2013.

Immune Effects

The committee conducted two literature searches to identify recent studies pertaining to immune effects after exposure to formaldehyde (see Box D-4). The first search resulted in 2,405 publications. Through this approach, National

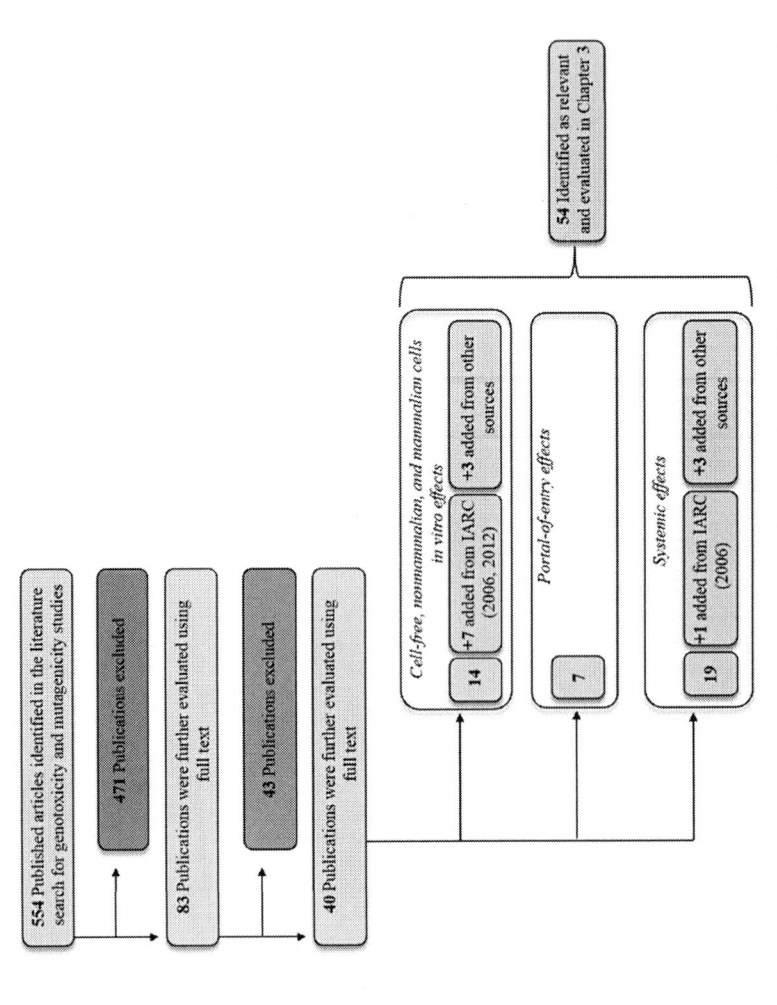

FIGURE D-3 Literature tree for genotoxicity search. See Box D-3 for a description of the exclusion criteria and search strategy.

Appendix D

> **BOX D-4** Exclusion Criteria and Search Strategy for Immune Effects
>
> Exclusion Criteria
>
> - The study did not evaluate health effects of formaldehyde or its metabolites known to be formed in humans.
> - The study evaluated immune effects not relevant to carcinogenesis.
> - The publication did not contain primary data.
>
> First Search Strategy
>
> PubMed: [("Formaldehyde"[Title/Abstract]) AND ("immun*" OR "bone marrow" OR "bone marrow"[MeSH] OR "lymphocytes" OR "lymphocytes"[MeSH] OR "hematopoietic" OR "allergy" OR "sensitization" OR "lymph node" OR leukopenia OR lymphocytopenia OR immunotoxicity)]. Search run on 05-10-2013 and updated on 11-08-2013; limited to 2009–2013.
>
> *Medline and Embase:* [(Formaldehyde.ab OR formaldehyde.ti) AND (immune*.mp OR bone marrow.mp. OR bone marrow/ OR lymphocytes.mp. OR lymphocyte/ OR hematopoietic.mp. OR allergy.mp. OR sensitization.mp. OR lymph node.mp. OR leucopenia.mp. OR lymphocytopenia.mp. OR immunotoxicity.mp.)]. Search run on 05-10-2013 and updated on 11-08-2013; limited to 2009–2013.
>
> *Scopus:* [("Formaldehyde") AND ("immun*" OR "bone marrow" OR "lymphocytes" OR "hematopoitic" OR "allergy" OR "sensitization" OR "lymph node"OR "leucopenia" OR "lymphocytopenia" OR "immunotoxicity")]. Search run on 05-10-2013 and updated on 11-08-2013; limited to 2009–2013.
>
> *Web of Science:* [("formaldehyde") AND ("immun*" OR "bone marrow" OR lymphocytes" OR "hematopoietic" OR "allergy" OR "sensitization" OR "lymph node"OR "leucopenia" OR "lymphocytopenia" OR "immunotoxicity")]. Search run on 05-10-2013 and updated on 11-08-2013; limited to 2009–2013.
>
> Second Search Strategy
>
> PubMed: [("Formaldehyde"[Title])]. Search run on 11-06-2013; limited to 2009–2013.

Research Council staff identified 46 studies that contributed an understanding of hematologic effects related to formaldehyde exposure of humans, animals, and isolated hematologic cell types (see Table 3-18). A committee member reviewed the abstracts in greater detail and identified 18 that warranted inclusion in the "Hematologic Effects" section of Chapter 3. To identify studies that may have been missed, a second search was performed with the search term "Formaldehyde[Title]" in Pubmed. There were 730 studies returned from the second search. Titles were reviewed to identify new studies not previously

206 *Review of the Formaldehyde Assessment in the NTP 12th Report on Carcinogens*

considered and, when appropriate, abstracts and full text were reviewed. The search resulted in identification of four additional studies. In its reading of the literature, the committee also identified three studies that were relevant to this section that were not cited in the background document or substance profile for formaldehyde. Those results are depicted in Figure D-4.

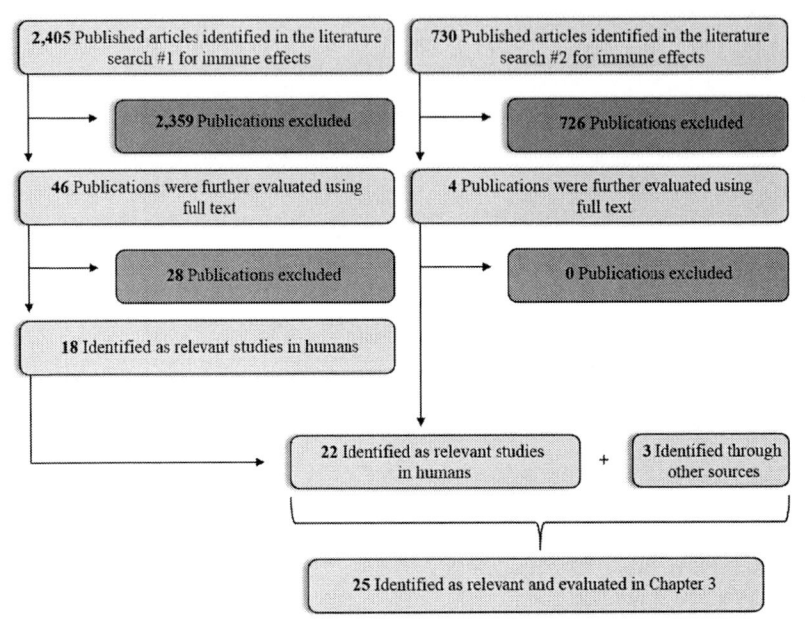

```
┌─────────────────────────────────┐         ┌─────────────────────────────────┐
│ 2,405 Published articles        │         │ 730 Published articles          │
│ identified in the literature    │         │ identified in the literature    │
│ search #1 for immune effects    │         │ search #2 for immune effects    │
└─────────────────────────────────┘         └─────────────────────────────────┘
         │                                           │
         ▼                                           ▼
  ┌──────────────────────┐                    ┌──────────────────────┐
  │ 2,359 Publications   │                    │ 726 Publications     │
  │ excluded             │                    │ excluded             │
  └──────────────────────┘                    └──────────────────────┘
         │                                           │
         ▼                                           ▼
┌─────────────────────────────────┐         ┌─────────────────────────────────┐
│ 46 Publications were further    │         │ 4 Publications were further     │
│ evaluated using full text       │         │ evaluated using full text       │
└─────────────────────────────────┘         └─────────────────────────────────┘
         │                                           │
         ▼                                           ▼
  ┌──────────────────────┐                    ┌──────────────────────┐
  │ 28 Publications      │                    │ 0 Publications       │
  │ excluded             │                    │ excluded             │
  └──────────────────────┘                    └──────────────────────┘
         │
         ▼
┌─────────────────────────────────┐
│ 18 Identified as relevant       │
│ studies in humans               │
└─────────────────────────────────┘
                           │
                           ▼
              ┌──────────────────────────┐    ┌──────────────────────┐
              │ 22 Identified as         │ +  │ 3 Identified through │
              │ relevant studies in      │    │ other sources        │
              │ humans                   │    │                      │
              └──────────────────────────┘    └──────────────────────┘
                                    │
                                    ▼
                    ┌──────────────────────────────────────────────┐
                    │ 25 Identified as relevant and evaluated in   │
                    │ Chapter 3                                    │
                    └──────────────────────────────────────────────┘
```

FIGURE D-4 Literature tree for immune-effects search. See Box D-4 for a description of the exclusion criteria and search strategy.

REFERENCES

Bucher, J.R. 2013. Follow-up Questions. Material submitted by the NAS Committee on Review of the Formaldehyde Assessment in the NTP 12th RoC and the NAS Committee on Review of the Styrene Assessment in the NTP 12th RoC, April 2, 2013.

Kushman, M.E., A.D. Kraft, K.Z. Guyton, W.A. Chiu, S.L. Makris, and I. Rusyn. 2013. A systematic approach for identifying and presenting mechanistic evidence in human health assessments. Regul. Toxicol. Pharmacol. 67(2):266-277

Appendix E

Genotoxicity and Mutagenicity Summary Tables

The committee undertook a comprehensive review of scientific peer-reviewed literature on formaldehyde genotoxicity and mutagenicity. The review included studies that were available to the National Toxicology Program at the time the 12th Report on Carcinogens was published and new literature published since July 10, 2011 (see the description of the literature search strategy, including the dates of the search, in Appendix D). The tables in this appendix provide information on the following outcomes: DNA adducts (Table E-1), DNA–DNA cross-links (Table E-2), DNA–protein cross-links (Table E-3), DNA strand breaks (Table E-4), mutations (Table E-5), sister-chromatid exchanges (Table E-6), micronuclei (Table E-7), and chromosomal aberrations (Table E-8). The evidence is organized by cell-free systems; nonmammalian model organisms; mammalian in vitro systems in the rodent, primate, and human; mammalian in vivo systems showing portal-of-entry effects in the rodent, primate, and human; and mammalian in vivo systems showing systemic effects in the rodent, primate, and human. The studies are categorized as either positive (the effect studied was statistically significant for the outcome of interest) or negative (the effect was studied, but no statistically significant change in the outcome of interest was observed). Table 3-9 summarizes the evidence.

TABLE E-1 DNA Adducts

		Positive Studies	Negative Studies
Cell-free systems		Von Hippel and Wong 1971[1] Beland et al. 1984[1] Snyder and Van Houten 1986[1] Zhong and Que Hee 2004a, 2005[1] Cheng et al. 2008[1] Lu et al. 2009[1]	—
Nonmammalian model organisms		—	—
Mammalian in vitro	Rodent	Beland et al. 1984[1]	—
	Human	Zhong and Que Hee 2004b[1] Lu et al. 2012[2]	—
Mammalian in vivo: portal-of-entry effects	Rodent	Lu et al. 2010a[1] Lu et al. 2011[2]	—
	Primate	Moeller et al. 2011[2]	—
	Human	—	—
Mammalian in vivo: systemic effects*	Rodent	—	Lu et al. 2010a[1]
	Primate	—	Moeller et al. 2011[2]
	Human	Bono et al. 2010[2,#]	—

*The committee acknowledges that although most investigators consider the effects on circulating blood mononucleated cells systemic because cells for the analyses were collected from the systemic circulation, it is also plausible that these cells may have been exposed to formaldehyde in the nose through lymphoid tissue in the mucosa.
#M1G adduct has been postulated to be the result of secondary DNA damage from formaldehyde-associated oxidative stress.
[1]The study was identified from the background document or the substance profile for formaldehyde in the National Toxicology Program 12th Report on Carcinogens (NTP 2010, 2011).
[2]The study was identified from the committee's new literature search (see Appendix D).
Source: Committee generated.

TABLE E-2 DNA–DNA Cross-Links

		Positive Studies	Negative Studies
Cell-free systems		Chaw et al. 1980[1] Huang et al. 1992[1] Huang and Hopkins 1993[1]	—
Nonmammalian model organisms		—	—
Mammalian in vitro	Rodent	—	—
	Human	—	—
Mammalian in vivo: portal-of-entry effects	Rodent	—	—
	Primate	—	—
	Human	—	—
Mammalian in vivo: systemic effects	Rodent	—	—
	Primate	—	—
	Human	—	—

[1]The study was identified from the background document or the substance profile for formaldehyde in the National Toxicology Program 12th Report on Carcinogens (NTP 2010, 2011).
Source: Committee generated.

TABLE E-3 DNA–Protein Cross-Links

		Positive Studies	Negative Studies
Cell-free systems		Kuykendall and Bogdanffy 1992[2] Lu et al. 2008[4] Lu et al. 2010b[4]	—
Nonmammalian model organisms		—	—
Mammalian in vitro	Rodent	Ross and Shipley 1980[1] Ross et al. 1981[1] Swenberg et al. 1983b[1] O'Connor and Fox 1987[1] Cosma et al. 1988a[1] Zhitkovich and Costa 1992[1] Olin et al. 1996[1] Casanova and Heck 1997[1] Casanova et al. 1997[1] Merk and Speit 1998, 1999[1] Speit et al. 2007a[1] Garcia et al. 2009[1] She et al. 2013[3]	Casanova et al. 1997[1]
	Human	Fornace et al. 1982[1] Grafström et al. 1984 Saladino et al. 1985[1] Grafström et al. 1986[1] Craft et al. 1987[1] Grafström 1990[1] Olin et al. 1996[1] Shaham et al. 1996a[1] Costa et al. 1997[1] Blasiak et al. 2000[1] Andersson et al. 2003[1] Emri et al. 2004[1]	—

Mammalian in vivo: portal-of-entry effects	Rodent	Casanova-Schmitz et al. 1984a[1] Lam et al. 1985[1] Casanova and Heck 1987[1] Heck et al. 1986, 1989[1] Cosma et al. 1988b[1] Casanova et al. 1989, 1994[1]	Saito et al. 2005[1] Liu et al. 2006[1] Schmid and Speit 2007[1] Speit et al. 2008b[1] Neuss et al. 2010a,b[3] Speit et al. 2010[3] Duan 2011[3] Zeller et al. 2011a[3] Wong et al. 2012[3] Ren et al. 2013[3]
	Primate	Heck et al. 1989[1] Casanova et al. 1991[1]	—
	Human	—	—
Mammalian in vivo: systemic effects	Rodent	Ke et al. 2012[3] Ye et al. 2013[3]	Casanova-Schmitz et al. 1984a[1] Casanova and Heck 1987[1]
	Primate	—	Heck et al. 1989[1] Casanova et al. 1991[1]
	Human	Shaham et al. 1996a, 2003[1] Lin et al. 2013[3]	—

[1]The study was identified from the background document or the substance profile for formaldehyde in the National Toxicology Program 12th Report on Carcinogens (NTP 2010, 2011).
[2]The study was identified from IARC 2006 or IARC 2012.
[3]The study was identified from the committee's new literature search (see Appendix D).
[4]The study was identified through additional ad hoc searches or from the reference list of other studies.
Source: Committee generated.

TABLE E-4 DNA Strand Breaks

		Positive Studies	Negative Studies
Cell-free systems		—	—
Nonmammalian model organisms		Poverenny et al. 1975[2] Wilkins and Macleod 1976[2] Magana-Schwencke et al. 1978[1] Magana-Schwencke and Ekert 1978[2] Magana-Schwencke and Moustacchi 1980[2] Le Curieux et al. 1993[2]	—
Mammalian in vitro	Rodent	Ross and Shipley 1980[1] O'Connor and Fox 1987[1] Cosma et al. 1988a[1] Demkowicz-Dobrzanski and Castonguay 1992[1] Kumari et al. 2012[3] She et al. 2013[3]	Ross et al. 1981[1] Speit et al. 2007a[1]
	Human	Fornace et al. 1982[1] Grafström et al. 1984[1] Saladino et al. 1985[1] Grafström et al. 1986[1] Snyder and Van Houten 1986[1] Grafström 1990[1] Vock et al. 1999[1] Liu et al. 2006[1]	—
Mammalian in vivo: portal-of-entry effects	Rodent	—	Neuss et al. 2010c[3]
	Primate	—	—
	Human	—	—

Mammalian in vivo: systemic effects	Rodent	Im et al. 2006[1] Wang and Liu 2006[1]	Speit et al. 2009[1]
	Primate	—	—
	Human	Yu et al. 2005[1] Jiang et al. 2006[1] Jiang et al. 2006[3] Tong et al. 2006[1] Costa et al. 2008[1] Jiang et al. 2010[3] Costa et al. 2011[4] Gomaa et al. 2012[3] Lin et al. 2013[3]	Zeller et al. 2011b[3] Aydin et al. 2013[4]

[1]The study was identified from the background document or the substance profile for formaldehyde in the National Toxicology Program 12th Report on Carcinogens (NTP 2010, 2011).
[2]The study was identified from IARC 2006 or IARC 2012.
[3]The study was identified from the committee's new literature search (see Appendix D).
[4]The study was identified through additional ad hoc searches or from the reference list of other studies.
Source: committee-generated.

TABLE E-5 Mutations

	Positive Studies	Negative Studies
Cell-free systems	—	—
Nonmammalian model organisms	Reviewed in IARC (2006)[2]: largely positive (with and without S9) for point mutations in bacteria (*Salmonella typhimurium*, *Escherichia coli*) and nonmammalian eukaryotes (*Neurospora crassa*, *Drosophila melanogaster*, *Caenorhabditis elegans*)	Reviewed in IARC (2006)[2]: largely negative for frame-shift mutations in *S. typhimurium*
Mammalian in vitro — Rodent	Grafstrom et al. 1993[1] Mackerer et al. 1996[1] Speit and Merk 2002[1]	Merk and Speit 1998, 1999[1]
Mammalian in vitro — Human	Goldmacher and Thilly 1983[1] Grafström et al. 1985[1] Craft et al. 1987[1] Crosby et al. 1988[1] Liber et al. 1989[1] Grafström 1990[1]	—
Mammalian in vivo: portal-of-entry effects — Rodent	Recio et al. 1992[1]	Meng et al. 2010[1]
Mammalian in vivo: portal-of-entry effects — Primate	—	—
Mammalian in vivo: portal-of-entry effects — Human	—	—
Mammalian in vivo: systemic effects — Rodent	Liu et al. 2009b[1]	—
Mammalian in vivo: systemic effects — Primate	—	—
Mammalian in vivo: systemic effects — Human	—	—

[1]The study was identified from the background document or the substance profile for formaldehyde in the National Toxicology Program 12th Report on Carcinogens (NTP 2010, 2011).
[2]The study was identified from IARC 2006 or IARC 2012.
Source: Committee generated.

TABLE E-6 Sister-Chromatid Exchanges

		Positive Studies	Negative Studies
Cell-free systems		—	—
Nonmammalian model organisms		—	—
Mammalian in vitro	Rodent	Obe and Beek 1979[1] Natarajan et al. 1983[1] Basler et al. 1985[1] Galloway et al. 1985[1] Merk and Speit 1998, 1999[1] Speit et al. 2007a[1] Garcia et al. 2009[1] She et al. 2013[2]	—
	Human	Obe and Beek 1979[1] Kreiger and Garry 1983[1] Schmid et al. 1986[1] Schmid and Speit 2007[1] Neuss and Speit 2008 Zeller et al. 2011a[2]	—
Mammalian in vivo: portal-of-entry effects	Rodent	—	—
	Primate	—	—
	Human	—	—
Mammalian in vivo: systemic effects	Rodent	—	Kligerman et al. 1984[1] Speit et al. 2009[1]
	Primate	—	—
	Human	Yager et al. 1986[1] Shaham et al. 1997, 2002[1] He et al. 1998[1]	Thomson et al. 1984[1] Bauchinger and Schmid 1985[1] Chebotarev et al. 1986[1] Suruda et al. 1993[1]

(Continued)

TABLE E-6 Continued

Positive Studies	Negative Studies
Ye et al. 2005[1]	Ying et al. 1999[1]
Costa et al. 2008[1]	Pala et al. 2008[1]
Costa et al. 2013[2]	Jakab et al. 2010[2]
	Zeller et al. 2011b, 2012[2]

[1]The study was identified from the background document or the substance profile for formaldehyde in the National Toxicology Program 12th Report on Carcinogens (NTP 2010, 2011).
[2]The study was identified from the committee's new literature search (see Appendix D).
Source: committee-generated.

TABLE E-7 Micronuclei

		Positive Studies	Negative Studies
Cell-free systems		—	—
Nonmammalian model organisms		—	—
Mammalian in vitro	Rodent	Merk and Speit 1998[1] Speit et al. 2007a[1] Ji et al. 2013[3] She et al. 2013[3]	
	Human	Speit et al. 2000[1] Schmid and Speit 2007[1] Speit et al. 2011a[3] Ren et al. 2013[3]	
Mammalian in vivo: portal-of-entry effects	Rodent	Migliore et al. 1989[1]	Neuss et al. 2010c[3] Speit et al. 2011b[3]
	Primate		
	Human	Ballarin et al. 1992[1] Suruda et al. 1993[1] Kitaeva et al. 1996[1] Titenko-Holland et al. 1996[1] Ying et al. 1997[1] Burgaz et al. 2001, 2002[1] Ye et al. 2005[1] Ladeira et al. 2011, 2013[3] Viegas et al. 2013[3]	Titenko-Holland et al. 1996[1] Speit et al. 2007b[1] Zeller et al. 2011b[3]
Mammalian in vivo: systemic effects	Rodent	Zhao et al. 2004[4] Gao et al. 2008[4] Gao et al. 2009[3] Katsnelson et al. 2013[4]	Gocke et al. 1981[1] Natarajan et al. 1983[1] Kim et al. 1991[4] Morita et al. 1997[1] Speit et al. 2009[1]

(Continued)

TABLE E-7 Continued

	Positive Studies	Negative Studies
Primate	—	—
Human	Suruda et al. 1993[1]	Ying et al. 1997[1]
	Kitaeva et al. 1996[1]	Pala et al. 2008[1]
	He et al. 1998[1]	Zeller et al. 2011b[3]
	Yu et al. 2005[2]	
	Orsiere et al. 2006[1]	
	Iarmarcovai et al. 2007[1]	
	Costa et al. 2008[1]	
	Jiang et al. 2010[3]	
	Viegas et al. 2010[4]	
	Brahem et al. 2011[3]	
	Costa et al. 2011[4]	
	Ladeira et al. 2011, 2013[3]	
	Bouraoui et al. 2013[3]	
	Costa et al. 2013[3]	
	Lin et al. 2013[3]	
	Viegas et al. 2013[3]	
	Souza and Devi 2014[3]	

[1]The study was identified from the background document or the substance profile for formaldehyde in the National Toxicology Program 12th Report on Carcinogens (NTP 2010, 2011).
[2]The study was identified from IARC 2006 or IARC 2012.
[3]The study was identified from the committee's new literature search (see Appendix D).
[4]The study was identified through additional ad hoc searches or from the reference list of other studies.

Source: Committee-generated.

TABLE E-8 Chromosomal Aberrations

		Positive Studies	Negative Studies
Cell-free systems		—	—
Nonmammalian model organisms		—	—
Mammalian in vitro	Rodent	Ishidate et al. 1981[1] Natarajan et al. 1983[1] Galloway et al. 1985[1] Hikiba et al. 2005[1] Hagiwara et al. 2006[1]	
	Human	Miretskaya and Shvartsman 1982[1] Levy et al. 1983[1] Schmid et al. 1986[1] Dresp and Bauchinger 1988[1] Pongsavee 2011[2] Ren et al. 2013[2]	Kuehner et al. 2012[2] Ji et al. 2013[2]
Mammalian in vivo: portal-of-entry effects	Rodent	Dallas et al. 1992[1]	—
	Primate	—	—
	Human	—	—
Mammalian in vivo: systemic effects	Rodent	Kitaeva et al. 1990[1] Gomaa et al. 2012[2]	Fontignie-Heubrechts 1981[1] Natarajan et al. 1983[1] Kligerman et al. 1984[1] Dallas et al. 1992[1] Speit et al. 2009[1]
	Primate	—	
	Human	Bauchinger and Schmid 1985[1] Chebotarev et al. 1986[1] Kitaeva et al. 1996[1] He et al. 1998[1] Lazutka et al. 1999[1]	Fleig et al. 1982[1] Thomson et al. 1984[1] Vargova et al. 1992[1] Vasudeva and Anand 1996[1] Pala et al. 2008[1]

(Continued)

TABLE E-8 Continued

Positive Studies	Negative Studies
Neri et al. 2006[1]	
Jakab et al. 2010[2]	
Zhang et al. 2010b[1]	
Santovito et al. 2011[2]	
Gomaa et al. 2012[2]	
Musak et al. 2013[2]	

[1]The study was identified from the background document or the substance profile for formaldehyde in the National Toxicology Program 12th Report on Carcinogens (NTP 2010, 2011).
[2]The study was identified from the committee's new literature search (see Appendix D).
Source: committee-generated.

REFERENCES

Andersson, M., E. Agurell, H. Vaghef, G. Bolcsfoldi, and B. Hellman. 2003. Extended-term cultures of human T-lymphocytes and the comet assay: A useful combination when testing for genotoxicity in vitro? Mutat. Res. 540(1):43-55.

Aydin, S., H. Canpinar, U. Undeger, D. Güc, M. Çolakoğlu, A. Kars, and N. Başaran. 2013. Assessment of immunotoxicity and genotoxicity in workers exposed to low concentrations of formaldehyde. Arch. Toxicol. 87(1):145-153.

Ballarin, C., F. Sarto, L. Giacomelli, G.B. Bartolucci, and E. Clonfero. 1992. Micronucleated cells in nasal mucosa of formaldehyde-exposed workers. Mutat. Res. 280(1):1-7.

Basler, A., W. Hude, and M. Scheutwinkel-Reich. 1985. Formaldehyde-induced sister chromatid exchanges in vitro and the influence of the exogenous metabolizing systems S9 mix and primary rat hepatocytes. Arch. Toxicol. 58(1):10-13.

Bauchinger, M., and E. Schmid. 1985. Citogenetic effects in lymphocytes of formaldehyde workers of a paper factory. Mutat. Res.158(3):195-199.

Beland, F.A., N.F. Fullerton, and R.H. Heflich. 1984. Rapid isolation, hydrolysis and chromatography of formaldehyde-modified DNA. J. Chromatogr. 308(6):121-131.

Blasiak, J., A. Trzeciak, E. Malecka-Panas, J. Drzewoski, and M. Wojewodzka. 2000. In vitro genotoxicity of ethanol and acetaldehyde in human lymphocytes and the gastrointestinal tract mucosa cells. Toxicol. In Vitro 14(4):287-295.

Bono, R., V. Romanazzi, A. Munnia, S. Piro, A. Allione, F. Ricceri, S. Guarrera, C. Pignata, G. Matullo, P. Wang, R.W. Giese, and M. Peluso. 2010. Malondialdehyde-deoxyguanosine adduct formation in workers of pathology wards: The role of air formaldehyde exposure. Chem. Res. Toxicol. 23(8):1342-1348.

Bouraoui, S., S. Mougou, A. Brahem, F. Tabka, H. Ben Khelifa, I Harrabi, N. Mrizek, H. Elghezal, and A. Saad. 2013. A combination of micronucleus assay and fluorescence in situ hybridization analysis to evaluate the genotoxicity of formaldehyde. Arch. Environ. Contam. Toxicol. 64(2):337-344.

Brahem, A., S. Bouraoui, H. ElGhazel, A. Ben Amor, A. Saad, F. Dabbebi, and N. Mrizek. 2011. Genotoxic risk assessment in an anatomic pathology laboratory by use of the micronucleus test [in French]. Arch. Mal. Prof. Environ. 72(4):370-375.

Burgaz, S., G. Cakmak, O. Erdem, M. Yilmaz, and A.E. Karakaya. 2001. Micronuclei frequencies in exfoliated nasal mucosa cells from pathology and anatomy laboratory workers exposed to formaldehyde. Neoplasma 48(2):144-147.

Burgaz, S., O. Erdem, G. Cakmak, N. Erdem, A. Karakaya, and A.E. Karakaya. 2002. Cytogenetic analysis of buccal cells from shoe-workers and pathology and anatomy laboratory workers exposed to n-hexane, toluene, methyl ethyl ketone and formaldehyde. Biomarkers 7(2):151-161.

Casanova, M., and H.D. Heck. 1987. Further studies of the metabolic incorporation and covalent binding of inhaled [3H]- and [14C]formaldehyde in Fischer-344 rats: Effects of glutathione depletion. Toxicol. Appl. Pharmacol. 89(1):105-121.

Casanova, M., and H.D. Heck. 1997. Lack of evidence for the involvement of formaldehyde in the hepatocarcinogenicity of methyl tertiary-butyl ether in CD-1 mice. Chem. Biol. Interact. 105(2):131-143.

Casanova, M., D.F. Deyo, and H.D. Heck. 1989. Covalent binding of inhaled formaldehyde to DNA in the nasal mucosa of Fisher 344 rats: Analysis of formaldehyde and DNA by high-performance liquid chromatography and provisional pharmacokinetic interpretation. Fundam. Appl. Toxicol. 12(3):397-417.

Casanova, M., K.T. Morgan, W.H. Steinhagen, J.I. Everitt, J.A. Popp, and H.D. Heck. 1991. Covalent binding of inhaled formaldehyde to DNA in the respiratory tract of rhesus monkey: Pharmacokinetics, rat to monkey interspecies scaling, and extrapolation to man. Fundam. Appl. Toxicol. 17(2):409-428.

Casanova, M., K.T. Morgan, E.A. Gross, O.R. Moss, and H.D. Heck. 1994. DNA-protein cross-links and cell replication at specific sites in the nose of F344 rats exposed subchronically to formaldehyde. Fundam. Appl. Toxicol. 23(4):525-536.

Casanova, M., D.A. Bell, and H.D. Heck. 1997. Dichloromethane metabolism to formaldehyde and reaction of formaldehyde with nucleic acids in hepatocytes of rodents and humans with and without glutathione S-transferase T1 and M1 genes. Fundam. Appl. Toxicol. 37(2):168-180.

Casanova-Schmitz, M., T.B. Starr, and H.D. Heck. 1984a. Differentiation between metabolic incorporation and covalent binding in the labeling of macromolecules in the rat nasal mucosa and bone marrow by inhaled [14C]- and [3H]formaldehyde. Toxicol. Appl. Pharmacol. 76(1):26-44.

Chaw, Y.F., L.E. Crane, P. Lange, and R. Shapiro. 1980. Isolation and identification of cross-links from formaldehyde-treated nucleic acids. Biochemistry 19(24):5525-5531.

Chebotarev, A.N., N.V. Titenko, T.G. Selezneva, V.N. Fomenko, and L.M. Katosova. 1986. Comparison of chromosome aberrations, sister chromatid exchanges and unscheduled DNA synthesis in the evaluation of the mutagenicity of environmental factors [in Russian]. Tsitol. Genet. 20(2):109-115.

Cheng, G., M. Wang, P. Upadhyaya, P.W. Villalta, and S.S. Hecht. 2008. Formation of formaldehyde adducts in the reactions of DNA and deoxyribonucleosides with alpha-acetates of 4-(methylnitrosamino)-1-(3-pyridyl)-1-butanone (NNK), 4-(methylnitrosamino)-1-(3-pyridyl)-1-butanol (NNAL), and N-nitrosodimethylamine (NDMA). Chem. Res. Toxicol. 21(3):746-751.

Cosma, G.N., R. Jamasbi, and A.C. Marchok. 1988a. Growth inhibition and DNA damage induced by benzo[a]pyrene and formaldehyde in primary cultures of rat tracheal epithelial cells. Mutat. Res. 201(1):161-168.

Cosma, G.N., A.S. Wilhite, and A.C. Marchok. 1988b. The detection of DNA-protein cross-links in rat tracheal implants exposed in vivo to benzo[a]pyrene and formaldehyde. Cancer Lett. 42(1-2):13-21.

Costa, M., A. Zhitkovich, M. Harris, D. Paustenbach, and M. Gargas. 1997. DNA-protein cross-links produced by various chemicals in cultured human lymphoma cells. J. Toxicol. Environ. Health 50(5):433-449.

Costa, S, P. Coelho, C. Costa, S. Silva, O. Mayan, L.S. Santos, J. Gaspar, and J.P. Teixeira. 2008. Genotoxic damage in pathology anatomy laboratory workers exposed to formaldehyde. Toxicology 252(1-3):40-48.

Costa, S., C. Pina, P. Coelho, C. Costa, S. Silva, B. Porto, B. Laffon, and J.P. Teixeira. 2011. Occupational exposure to formaldehyde: Genotoxic risk evaluation by comet assay and micronucleus test using human peripheral lymphocytes. J. Toxicol. Environ. Health A 74(15-16):1040-1051.

Costa, S., J. Garcia-Leston, M. Coelho, P. Coelho, C. Costa, S. Silva, B. Porto, B. Laffon, and J.P. Teixeira. 2013. Cytogenetic and immunological effects associated with occupational formaldehyde exposure. J. Toxicol. Environ. Health A 76(4-5):217-229.

Craft, T.R., E. Bermudez, and T.R. Skopek. 1987. Formaldehyde mutagenesis and formation of DNA-protein crosslinks in human lymphoblasts in vitro. Mutat. Res. 176(1):147-155.

Crosby, R.M., K.K. Richardson, T.R. Craft, K.B. Benforado, H.L. Liber, and T.R. Skopek. 1988. Molecular analysis of formaldehyde-induced mutations in human lymphoblasts and E. coli. Environ. Mol. Mutagen. 12(2):155-166.

Dallas, C.E., M.J. Scott, J.B. Ward, Jr., and J.C. Theiss. 1992. Cytogenetic analysis of pulmonary lavage and bone marrow cells of rats after repeated formaldehyde inhalation. J. Appl. Toxicol. 12(3):199-203.

Demkowicz-Dobrzanski, K., and A. Castonguay. 1992. Modulation by glutathione of DNA strand breaks induced by-(methylnitrosamino)-1-(3-pyridyl)-1-butanone and its aldehyde metabolites in rat hepatocytes. Carcinogenesis 13(8):1447-1454.

Dresp, J., and M. Bauchinger. 1988. Direct analysis of the clastogenic effect of formaldehyde in unstimulated human lymphocytes by means of the premature chromosome condensation technique. Mutat. Res. 204(2):349-352.

Duan, Y.Y. 2011. Effects of overexpression of heat shock protein 70 on the damage induced by formaldehyde in vitro [in Chinese]. Zhonghua Lao Dong Wei Sheng Zhi Ye Bing Za Zhi 29(5):349-352.

Emri, G., D. Schaefer, B. Held, C. Herbst, W. Zieger, I. Horkay, and C. Bayerl. 2004. Low concentrations of formaldehyde induce DNA damage and delay DNA repair after UV irradiation in human skin cells. Exp. Dermatol. 13(5):305-315.

Fleig, I., N. Petri, W.G. Stocker, and A.M. Thiess. 1982. Cytogenetic analyses of blood lymphocytes of workers exposed to formaldehyde in formaldehyde manufacturing and processing. J. Occup. Environ. Med. 24(12):1009-1012.

Fontignie-Houbrechts, N. 1981. Genetic effects of formaldehyde in the mouse. Mutat. Res. 88(1):109-114.

Fornace, A.J., J.F. Lechner, R.C. Grafstrom, and C.C. Harris. 1982. DNA repair in human bronchial epithelial cells. Carcinogenesis 3(12):1373-1377.

Galloway, S.M., A.D. Bloom, M. Resnick, B.H. Margolin, F. Nakamura, P. Archer, and E. Zeiger. 1985. Development of a standard protocol for in vitro cytogenetic testing with Chinese hamster ovary cells: Comparison of results for 22 compounds in two laboratories. Environ. Mutagen. 7(1):1-51.

Gao, N., Q. Chang, L. Chen, X. Li, X. Yang, and S. Ding. 2008. Effect of micronucleus rate in peripheral lymphocytes and liver cells of mice exposed to gaseous formaldehyde [in Chinese]. J. Public Health Prev. Med. 19(3):7-9.

Gao, N., Q. Chang, D. Liu, W. Cheng, and S. Ding. 2009. Study on genotoxicity of formaldehyde on the liver cells in mice. ICBBE 2009: The Third International Conference on Bioinformatics and Biomedical Engineering, June 11-13, 2009, Beijing, China [online]. Available: http://ieeexplore.ieee.org/xpls/abs_all.jsp?arnumber=5163474 [accessed July 8, 2014].

Garcia, C.L., M. Mechilli, L.P. De Santis, A. Schinoppi, K. Kobos, and F. Palitti. 2009. Relationship between DNA lesions, DNA repair and chromosomal damage induced by acetaldehyde. Mutat. Res. 662(1-2):3-9.

Gocke, E., M.T. King, K. Eckhardt, and D. Wild. 1981. Mutagenicity of cosmetics ingredients licensed by the European Communities. Mutat. Res. 90(2):91-109.

Goldmacher, V.S., and W.G. Thilly. 1983. Formaldehyde is mutagenic for cultured human cells. Mutat. Res. 116(3-4):417-422.

Gomaa, M.S., G.E. Elmesallamy, and M.M. Sameer. 2012. Evaluation of genotoxic effects of formaldehyde in adult albino rats and its implication in case of human exposure. Life Sci. J. 4(9):3085-3093.

Grafström, R.C. 1990. In vitro studies of aldehyde effects related to human respiratory carcinogenesis. Mutat. Res. 238(3):175-184.

Grafström, R.C., A. Fornace, and C.C. Harris. 1984. Repair of DNA damage caused by formaldehyde in human cells. Cancer Res. 44(10):4323-4327.

Grafström, R.C., R.D. Curren, L.L. Yang, and C.C. Harris. 1985. Genotoxicity of formaldehyde in cultured human bronchial fibroblasts. Science 228(4695):89-91.

Grafström, R.C., J.C. Wiley, K. Sundqvist, and C.C. Harris. 1986. Pathobiological effects of tobacco smoke-related aldehydes in cultured human bronchial epithelial cells. Pp. 273-285 in Mechanisms in Tobacco Carcinogenesis, D. Hoffman, and C.C. Harris, eds. Banbury Report 23. Cold Spring Harbor, NY: CSH Press.

Grafström, R.C., I.C. Hsu, and C.C. Harris. 1993. Mutagenicity of formaldehyde in Chinese hamster lung fibroblasts: Synergy with ionizing radiation and N-nitroso-N-methylurea. Chem. Biol. Interact. 86(1):41-49.

Hagiwara, M., E. Watanabe, J.C. Barrett, and T. Tsutsui. 2006. Assessment of genotoxicity of 14 chemical agents used in dental practice: Ability to induce chromosome aberrations in Syrian hamster embryo cells. Mutat. Res. 603(2):111-120.

He, J.L., L.F. Jin, and H.Y. Jin. 1998. Detection of cytogenetic effects in peripheral lymphocytes of students exposed to formaldehyde with cytokinesis-blocked micronucleus assay. Biomed. Environ. Sci. 11(1):87-92.

Heck, H.A, M. Casanova, C.W. Lam, and J.A. Swenberg. 1986. The formation of DNA-protein cross-links by aldehydes present in tobacco smoke. Pp. 215-230 in Mechanisms in Tobacco Carcinogenesis, D. Hoffman, and C.C. Harris, eds. Banbury Report 23. Cold Spring Harbor, NY: CSH Press.

Heck, H.A., M. Casanova, W.H. Steinhagen, J.I. Everitt, K.T. Morgan, and J.A. Popp. 1989. Formaldehyde toxicity: DNA-protein cross-linking studies in rats and non-human primates. Pp. 159-164 in Nasal Carcinogenesis in Rodents: Relevance to Human Risk, V.J. Feron, and M.C. Bosland, eds. Wageningen: Pudoc.

Hikiba, H., E. Watanabe, J.C. Barrett, and T. Tsutsui. 2005. Ability of fourteen chemical agents used in dental practice to induce chromosome aberrations in Syrian hamster embryo cells. J. Pharmacol. Sci. 97(1):146-152.

Huang, H., and P.B. Hopkins. 1993. DNA interstrand cross-linking by formaldehyde: Nucleotide sequence preference and covalent structure of the predominant cross-link formed in synthetic oligonucleotides. J. Am. Chem. Soc. 115(21):9402-9408.

Huang, H.F., M.S. Solomon, and P.B. Hopkins. 1992. Formaldehyde preferentially interstrand cross-links duplex DNA through deoxyadenosine residues at the sequence 5'-d(AT). J. Am. Chem. Soc. 114(23):9240-9241.

IARC (International Agency for Research on Cancer). 2006. Formaldehyde. Pp. 39-325 in Formaldehyde, 2-Butoxyethanol and 1-tert-Butoxypropan-2-ol. IARC Monographs on the Evaluation of the Carcinogenic Risks to Humans Vol. 88. Lyon, France: IARC [online]. Available: http://monographs.iarc.fr/ENG/Monographs/vol88/mono88.pdf [accessed Jan. 10, 2014].

IARC (International Agency for Research on Cancer). 2012. Formaldehyde. Pp.401-436 in Review of Human Carcinogens: Chemical Agents and Related Occupations. IARC Monographs on the Evaluation of the Carcinogenic Risks to Humans Vol. 100F. Lyon, France: IARC [online]. Available: http://monographs.iarc.fr/ENG/Monographs/vol100F/mono100F.pdf [accessed Jan. 27, 2014].

Iarmarcovai, G., S. Bonassi, I. Sari-Minodier, M. Baciuchka-Palmaro, A. Botta, and T. Orsiere. 2007. Exposure to genotoxic agents, host factors, and lifestyle influence the number of centromeric signals in micronuclei: A pooled re-analysis. Mutat Res. 615(1-2):18-27.

Im, H., E. Oh, J. Mun, J.Y. Khim, E. Lee, H.S. Kang, E. Kim, H. Kim, N.H. Won, Y.H. Kim, W.W. Jung, and D. Sul. 2006. Evaluation of toxicological monitoring mark-

ers using proteomic analysis in rats exposed to formaldehyde. J. Proteome Res. 5(6):1354-1366.
Ishidate, M., T. Sofuni, and K. Yoshikawa. 1981. Chromosomal aberration tests in vitro as a primary screening tool for environmental mutagens and/or carcinogens. Gann Monogr. Cancer Res. 27:95-108.
Jakab, M.G., T. Klupp, K. Besenyei, A. Biro, J. Major, and A. Tompa. 2010. Formaldehyde-induced chromosomal aberrations and apoptosis in peripheral blood lymphocytes of personnel working in pathology departments. Mutat. Res. 698(1-2):11-17.
Ji, Z., X. Li, M. Fromowitz, E. Mutter-Rottmayer, J. Tung, M. Smith, and L. Zhang. 2013. Formaldehyde induces micronuclei in mouse erythropoietic cells and suppresses the expansion of human erythroid progenitor cells. Toxicol. Lett. 224(2):233-239.
Jiang, S.F., L.Q. Yu, S.G. Leng, Y.S. Zhang, J. Cheng, Y.F. Dai, Y. Niu, F.S. He, and Y.X. Zheng. 2006. Association between XRCC1 gene polymorphisms and DNA damage of workers exposed to formaldehyde [in Chinese]. Wei Sheng Yan Jiu 35(6):675-677.
Jiang, S.F., L. Yu, J. Cheng, S. Leng, Y. Dai, Y. Zhang, Y. Niu, H. Yan, W. Qu, C. Zhang, K. Zhang, R. Yang, L. Zhou, and Y. Zheng. 2010. Genomic damages in peripheral blood lymphocytes and association with polymorphisms of three glutathione S-transferases in workers exposed to formaldehyde. Mutat. Res. 695(1-2):9-15.
Katsnelson, B.A., T.D. Degtyareva, L.I. Privalova, I.A. Minigaliyeva, T.V. Slyshkina, V.V. Ryzhov, and O.Y. Beresneva. 2013. Attenuation of subchronic formaldehyde inhalation toxicity with oral administration of glutamate, glycine and methionine. Toxicol. Lett. 220(2):181-186.
Ke, Y.J., X.D. Qin, L. Li, J. Du, Y.C. Zhang, and S.M. Ding. 2012. Toxic effect of formaldehyde on mouse bone marrow. China Environ. Sci. 32(6):1129-1133.
Kim, C.Y., K. Kim, J.S. Shim, Y.H. Kim, and J.K. Roh. 1991. Acute toxicity and micronucleus formation study in mice exposed to formaldehyde by inhalation [in Korean]. Korean J. Toxicol 7(1):61-71.
Kitaeva, L.V., E.M. Kitaev, and M.N. Pimenova. 1990. The cytopathic and cytogenetic sequelae of chronic inhalational exposure to formaldehyde on female germ cells and bone marrow cells in rats [in Russian]. Tsitologiia 32(12):1212-1216.
Kitaeva, L.V., E.A. Mikheeva, L.F. Shelomova, and P. Shvartsman. 1996. Genotoxic effect of formaldehyde in somatic human cells in vivo [in Russian]. Genetika 32(9):1287-1290.
Kligerman, A.D., M.C. Phelps, and G.L. Erexson. 1984. Cytogenetic analysis of lymphocytes from rats following formaldehyde inhalation. Toxicol. Lett. 21(3):241-246.
Kreiger, R.A., and V.F. Garry. 1983. Formaldehyde-induced cytotoxicity and sister-chromatid exchanges in human lymphocyte cultures. Mutat. Res. 120(1):51-55.
Kuehner, S., M. Schlaier, K. Schwarz, and G. Speit. 2012. Analysis of leukemia-specific aneuploidies in cultured myeloid progenitor cells in the absence and presence of formaldehyde exposure. Toxicol. Sci. 128(1):72-78.
Kumari, A., Y.X. Lim, A.H. Newell, S.B. Olson, and A.K. McCullough. 2012. Formaldehyde-induced genome instability is suppressed by an XPF-dependent pathway. DNA Repair 11(3):236-246.
Kuykendall, J.R., and M.S. Bogdanffy. 1992. Efficiency of DNA-histone crosslinking induced by saturated and unsaturated aldehydes in vitro. Mutat. Res. 283(2):131-136.

Ladeira, C., S. Viegas, E. Carolino, J. Prista, M.C. Gomes, and M. Brito. 2011. Genotoxicity biomarkers in occupational exposure to formaldehyde--the case of histopathology laboratories. Mutat. Res. 721(1):15-20.

Ladeira, C., S. Viegas, E. Carolino, M.C. Gomes, and M. Brito. 2013. The influence of genetic polymorphisms in XRCC3 and ADH5 genes on the frequency of genotoxicity biomarkers in workers exposed to formaldehyde. Environ. Mol. Mutagen. 54(3):213-221.

Lam, C.W., M. Casanova, and H.D. Heck. 1985. Depletion of nasal mucosal glutathione by acrolein and enhancement of formaldehyde-induced DNA-protein cross-linking by simultaneous exposure to acrolein. Arch. Toxicol. 58(2):67-71.

Lazutka, J.R., R. Lekevicius, V. Dedonyte, L. Maciuleviciute-Gervers, J. Mierauskiene, S. Rudaitiene, and G. Slapsyte. 1999. Chromosomal aberrations and sister-chromatid exchanges in Lithuanian populations: Effects of occupational and environmental exposures. Mutat. Res. 445(2):225-239.

Le Curieux, F., D. Marzin, and F. Erb. 1993. Comparison of three short-term assays: Results on seven chemicals. Potential contribution to the control of water genotoxicity. Mutat. Res. 319(3):223-236.

Levy, S., S. Nocentini, and C. Billardon. 1983. Induction of cytogenetic effects in human fibroblast cultures after exposure to formaldehyde or X-rays. Mutat. Res. 119(3):309-317.

Liber, H.L., K. Benforado, R.M. Crosby, D. Simpson, and T.R. Skopek. 1989. Formaldehyde-induced and spontaneous alterations in human hprt DNA sequence and mRNA expression. Mutat Res. 226(1):31-37.

Lin, D., Y. Guo, J. Yi, D. Kuang, X. Li, H. Deng, K. Huang, L. Guan, Y. He, X. Zhang, D. Hu, Z. Zhang, H. Zheng, X. Zhang, C.M. McHale, L. Zhang, and T. Wu. 2013. Occupational exposure to formaldehyde and genetic damage in the peripheral blood lymphocytes of plywood workers. J. Occup. Health 55(4):284-291.

Liu, Y.R., Y. Zhou, W. Qiu, J.Y. Zheng, L.L. Shen, A.P. Li, and J.W. Zhou. 2009b. Exposure to formaldehyde induces heritable DNA mutations in mice. J. Toxicol. Environ. Health A 72(11-12):767-773.

Liu, Y.S., C.M. Li, Z.S. Lu, S.M. Ding, X. Yang, and J.W. Mo. 2006. Studies on formation and repair of formaldehyde-damaged DNA by detection of DNA-protein crosslinks and DNA breaks. Front. Biosci. 11:991-997.

Lu, K., G. Boysen, L. Gao, L.B. Collins, and J.A. Swenberg. 2008. Formaldehyde-induced histone modifications in vitro. Chem. Res. Toxicol. 21(8):1586-1593.

Lu, K, W.J. Ye, A. Gold, L.M. Ball, J.A. Swenberg. 2009. Formation of S- 1-(N-2-deoxyguanosinyl)methyl glutathione between glutathione and DNA induced by formaldehyde. J. Am. Chem. Soc. 131(10):3414-3415.

Lu, K., L.B. Collins, H. Ru, E. Bermudez, and J.A. Swenberg. 2010a. Distribution of DNA adducts caused by inhaled formaldehyde is consistent with induction of nasal carcinoma but not leukemia. Toxicol. Sci. 116(2):441-451.

Lu, K., W.J. Ye, L. Zhou, L.B. Collins, X. Chen, A. Gold, L.M. Ball, and J.A. Swenberg. 2010b. Structural characterization of formaldehyde-induced cross-links between amino acids and deoxynucleosides and their oligomers. J. Am. Chem. Soc. 132(10):3388-3399.

Lu, K., B. Moeller, M. Doyle-Eisele, J. McDonald, J.A. Swenberg. 2011. Molecular dosimetry of N2-hydroxymethyl-dG DNA adducts in rats exposed to formaldehyde. Chem. Res. Toxicol. 24(2):159-161.

Lu, K., S. Craft, J. Nakamura, B.C. Moeller, and J.A. Swenberg. 2012. Use of LC-MS/MS and stable isotopes to differentiate hydroxymethyl and methyl DNA adducts from formaldehyde and nitrosodimethylamine. Chem. Res. Toxicol. 25(3):664-675.

Mackerer, C.R., F.A. Angelosanto, G.R. Blackburn, and C.A. Schreiner. 1996. Identification of formaldehyde as the metabolite responsible for the mutagenicity of methyl tertiary-butyl ether in the activated mouse lymphoma assay. Proc. Soc. Exp. Biol. Med. 212(4):338-341.

Magana-Schwencke, N., and B. Ekert. 1978. Biochemical analysis of damage induced in yeast by formaldehyde. 2. Induction of cross-links between DNA and protein. Mutat. Res. 51(1):11-19.

Magana-Schwencke, N., and E. Moustacchi. 1980. Biochemical analysis of damage induced in yeast by formaldehyde. 3. Repair of induced cross-links between DNA and proteins in the wild-type and in excision-deficient strains. Mutat. Res. 70(1):29-35.

Magana-Schwencke, N., B. Ekert, and E. Moustacchi. 1978. Biochemical analysis of damage induced in yeast by formaldehyde. 1. Induction of single-strand breaks in DNA and their repair. Mutat. Res. 50(2):181-193.

Meng, F., E. Bermudez, P.B. McKinzie, M.E. Andersen, H.J. Clewell, III, and B.L. Parsons. 2010. Measurement of tumor-associated mutations in the nasal mucosa of rats exposed to varying doses of formaldehyde. Regul. Toxicol. Pharmacol. 57(2-3):274-283.

Merk, O., and G. Speit. 1998. Significance of formaldehyde-induced DNA-protein cross-links for mutagenesis. Environ. Mol. Mutagen. 32(3):260-268.

Merk, O., and G. Speit. 1999. Detection of crosslinks with the comet assay in relationship to genotoxicity and cytotoxicity. Environ. Mol. Mutagen. 33(2):167-172.

Migliore, L., L. Ventura, R. Barale, N. Loprieno, S. Castellino, and R. Pulci. 1989. Micronuclei and nuclear anomalies induced in the gastro-intestinal epithelium of rats treated with formaldehyde. Mutagenesis 4(5):327-334.

Miretskaya, L.M., and P. Shvartsman. 1982. Chromosome damages in human lymphocytes as affected by formaldehyde. 1. Formaldehyde treatment of lymphocytes in culture [in Russian]. Tsitologiya 24(9):1056-1060.

Moeller, B.C., K. Lu, M. Doyle-Eisele, J. McDonald, A. Gigliotti, and J.A. Swenberg. 2011. Determination of N2-hydroxymethyl-dG adducts in the nasal epithelium and bone marrow of nonhuman primates following 13CD2-formaldehyde inhalation exposure. Chem. Res. Toxicol. 24(2):162-164.

Morita, T., N. Asano, T. Awogi, Y.F. Sasaki, S. Sato, H. Shimada, S. Sutou, T. Suzuki, A. Wakata, T. Sofuni, and M. Hayashi. 1997. Evaluation of the rodent micronucleus assay in the screening of IARC carcinogens (groups 1, 2A and 2B) the summary report of the 6th collaborative study by CSGMT/JEMS MMS. Collaborative Study of the Micronucleus Group Test. Mammalian Mutagenicity Study Group. Mutat. Res. 389(1):3-122.

Musak, L., Z. Smerhovsky, E. Halasova, O. Osina, L. Letkova, L. Vodickova, V. Polakova, J. Buchancova, K. Hemminki, and P. Vodicka. 2013. Chromosomal damage among medical staff occupationally exposed to volatile anesthetics, antineoplastic drugs, and formaldehyde. Scand. J. Work Environ. Health 39(6):618-630.

Natarajan, A.T., F. Darroudi, C.J. Bussman, and A.C. van Kesteren-van Leeuwen. 1983. Evaluation of the mutagenicity of formaldehyde in mammalian cytogenetic assays in vivo and vitro. Mutat. Res. 122(3-4):355-360.

Neri, M., S. Bonassi, L.E. Knudsen, R.J. Sram, N. Holland, D. Ugolini, and D.F. Merlo. 2006. Children's exposure to environmental pollutants and biomarkers of genetic damage. 1. Overview and critical issues. Mutat. Res. 612(1):1-13.

Neuss, S., and G. Speit. 2008. Further characterization of the genotoxicity of formaldehyde in vitro by the sister chromatid exchange test and co-cultivation experiments. Mutagenesis 23(5):355-357.

Neuss, S., B. Moepps, and G. Speit. 2010a. Exposure of human nasal epithelial cells to formaldehyde does not lead to DNA damage in lymphocytes after co-cultivation. Mutagenesis 25(4):359-364.

Neuss, S., K. Holzmann, and G. Speit. 2010b. Gene expression changes in primary human nasal epithelial cells exposed to formaldehyde in vitro. Toxicol. Lett. 198(2):289-295.

Neuss, S., J. Zeller, L. Ma-Hock, and G. Speit. 2010c. Inhalation of formaldehyde does not induce genotoxic effects in broncho-alveolar lavage (BAL) cells of rats. Mutat. Res. 695(1-2):61-68.

NTP (National Toxicology Program). 2010. Report on Carcinogens Background Document for Formaldehyde, January 22, 2010. U.S. Department of Health and Human Services, Public Health Service, National Toxicology Program, Research Triangle Park, NC [online]. Available: http://ntp.niehs.nih.gov/ntp/roc/twelfth/2009/November/Formaldehyde_BD_Final.pdf [accessed July 17, 2013].

NTP (National Toxicology Program). 2011. Formaldehyde. Pp. 195-205 in Report on Carcinogens, 12th Ed. U.S. Department of Health and Human Services, Public Health Service, National Toxicology Program, Research Triangle Park, NC [online]. Available: http://ntp.niehs.nih.gov/ntp/roc/twelfth/profiles/formaldehyde.pdf [accessed July 17, 2013].

Obe, G., and B. Beek. 1979. Mutagenic activity of aldehydes. Drug Alcohol Depend. 4(1-2):91-94.

O'Connor, P.M., and B.W. Fox. 1987. Comparative studies of DNA cross-linking reactions following methylene dimethanesulphonate and its hydrolytic product, formaldehyde. Cancer Chemother. Pharmacol. 19(1):11-15.

Olin, K.L., G.N. Cherr, E. Rifkin, and C.L. Keen. 1996. The effects of some redox-active metals and reactive aldehydes on DNA-protein cross-links in vitro. Toxicology 110(1-3):1-8.

Orsiere, T., I. Sari-Minodier, G. Iarmarcovai, and A. Botta. 2006. Genotoxic risk assessment of pathology and anatomy laboratory workers exposed to formaldehyde by use of personal air sampling and analysis of DNA damage in peripheral lymphocytes. Mutat. Res. 605(1-2):30-41.

Pala, M., D. Ugolini, M. Ceppi, F. Rizzo, L. Maiorana, C. Bolognesi, T. Schiliro, G. Gilli, P. Bigatti, R. Bono, and D. Vecchio. 2008. Occupational exposure to formaldehyde and biological monitoring of Research Institute workers. Cancer Detect. Prev. 32(2):121-126.

Pongsavee, M. 2011. In vitro study of lymphocyte antiproliferation and cytogenetic effect by occupational formaldehyde exposure. Toxicol. Ind. Health 27(8):719-723.

Poverenny, A.M., Y.A. Siomin, A.S. Saenko, and B.I. Sinzinis. 1975. Possible mechanisms of lethal and mutagenic action of formaldehyde. Mutat. Res. 27(1):123-126.

Recio, L., S. Sisk, L. Pluta, E. Bermudez, E.A. Gross, Z. Chen, K. Morgan, and C. Walker. 1992. p53 mutations in formaldehyde-induced nasal squamous cell carcinomas in rats. Cancer Res. 52(21):6113-6116.

Ren, X., Z. Ji, C.M. McHale, J. Yuh, J. Bersonda, M. Tang, M.T. Smith, and L. Zhang. 2013. The impact of FANCD2 deficiency on formaldehyde-induced toxicity in human lymphoblastoid cell lines. Arch. Toxicol. 87(1):189-196.

Ross, W.E., and N. Shipley. 1980. Relationship between DNA damage and survival in formaldehyde-treated mouse cells. Mutat. Res. 79(3):277-283.

Ross, W.E., D.R. McMillan, and C.F. Ross. 1981. Comparison of DNA damage by methylmelamines and formaldehyde. J. Natl. Cancer Inst. 67(1):217-221.

Saito, Y., K. Nishio, Y. Yoshida, and E. Niki. 2005. Cytotoxic effect of formaldehyde with free radicals via increment of cellular reactive oxygen species. Toxicology 210(2-3):235-245.

Saladino, A.J., J.C. Willey, J.F. Lechner, R.C. Grafstrom, M. LaVeck, and C.C. Harris. 1985. Effects of formaldehyde, acetaldehyde, benzoyl peroxide, and hydrogen peroxide on cultured normal human bronchial epithelial cells. Cancer Res. 45(6):2522-2526.

Santovito, A., T. Schiliro, S. Castellano, P. Cervella, M.P. Bigatti, G. Gilli, R. Bono, and M. DelPero. 2011. Combined analysis of chromosomal aberrations and glutathione S-transferase M1 and T1 polymorphisms in pathologists occupationally exposed to formaldehyde. Arch. Toxicol. 85(10):1295-1302.

Schmid, E., W. Goggelmann, and M. Bauchinger. 1986. Formaldehyde-induced cytotoxic, genotoxic and mutagenic response in human lymphocytes and Salmonella typhimurium. Mutagenesis 1(6):427-431.

Schmid, O., and G. Speit. 2007. Genotoxic effects induced by formaldehyde in human blood and implications for the interpretation of biomonitoring studies. Mutagenesis 22(1):69-74.

Shaham, J., Y. Bomstein, A. Meltzer, Z. Kaufman, E. Palma, and J. Ribak. 1996a. DNA - protein crosslinks, a biomarker of exposure to formaldehyde - in vitro and in vivo studies. Carcinogenesis 17(1):121-125.

Shaham, J., Y. Bomstein, A. Meltzer, and J. Ribak. 1997. DNA-protein crosslinks and sister chromatid exchanges as biomarkers of exposure to formaldehyde. Int. J. Occup. Environ. Health 3(2):95-104.

Shaham, J., R. Gurvich, and Z. Kaufman. 2002. Sister chromatid exchange in pathology staff occupationally exposed to formaldehyde. Mutat. Res. 514(1-2):115-123.

Shaham, J., Y. Bomstein, R. Gurvich, M. Rashkovsky, and Z. Kaufman. 2003. DNA-protein crosslinks and p53 protein expression in relation to occupational exposure to formaldehyde. Occup. Environ. Med. 60(6):403-409.

She, Y., Y. Li, Y. Liu, G. Asai, S. Sun, J. He, Z. Pan, and Y. Cui. 2013. Formaldehyde induces toxic effects and regulates the expression of damage response genes in BM-MSCs. Acta Biochim. Biophys. Sin. 45(12):1011-1020.

Snyder, R.D., and B. Van Houten. 1986. Genotoxicity of formaldehyde and an evaluation of its effects on the DNA repair process in human diploid fibroblasts. Mutat. Res. 165(1):21-30.

Souza, A., and R. Devi. 2014. Cytokinesis blocked micronucleus assay of peripheral lymphocytes revealing the genotoxic effect of formaldehyde exposure. Clin. Anat.27(3):308-312.

Speit, G., and O. Merk. 2002. Evaluation of mutagenic effects of formaldehyde in vitro: Detection of crosslinks and mutations in mouse lymphoma cells. Mutagenesis 17(3):183-187.

Speit, G., P. Schutz, and O. Merk. 2000. Induction and repair of formaldehyde-induced DNA-protein crosslinks in repair-deficient human cell lines. Mutagenesis 15(1):85-90.

Speit, G., P. Schutz, J. Hogel, and O. Schmid. 2007a. Characterization of the genotoxic potential of formaldehyde in V79 cells. Mutagenesis 22(6):387-394.

Speit, G., O. Schmid, M. Frohler-Keller, I. Lang, and G. Triebig. 2007b. Assessment of local genotoxic effects of formaldehyde in humans measured by the micronucleus test with exfoliated buccal mucosa cells. Mutat. Res. 627(2):129-135.

Speit, G., O. Schmid, S. Neuss, and P. Schutz. 2008b. Genotoxic effects of formaldehyde in the human lung cell line A549 and in primary human nasal epithelial cells. Environ. Mol. Mutagen. 49(4):300-307.

Speit, G., J. Zeller, O. Schmid, A. Elhajouji, L. Ma-Hock, and S. Neuss. 2009. Inhalation of formaldehyde does not induce systemic genotoxic effects in rats. Mutat. Res. 677(1-2):76-85.

Speit, G., S. Neuss, and O. Schmid. 2010. The human lung cell line A549 does not develop adaptive protection against the DNA-damaging action of formaldehyde. Environ. Mol. Mutagen. 51(2):130-137.

Speit, G., S. Kuhner, R. Linsenmeyer, and P. Schutz. 2011a. Does formaldehyde induce aneuploidy? Mutagenesis 26(6):805-811.

Speit, G., P. Schutz, I. Weber, I. Ma-Hock, W. Kaufmann, H.P. Gelbke, and S. Durrer. 2011b. Analysis of micronuclei, histopathological changes and cell proliferation in nasal epithelium cells of rats after exposure to formaldehyde by inhalation. Mutat. Res. 721(2):127-135.

Suruda, A., P. Schulte, M. Boeniger, R.B. Hayes, G.K. Livingston, K. Steenland, P. Stewart, R. Herrick, D. Douthit, and M.A. Fingerhut. 1993. Cytogenetic effects of formaldehyde exposure in students of mortuary science. Cancer Epidemiol. Biomarkers Prev. 2(5):453-460.

Swenberg, J.A., E.A. Gross, J. Martin, and J.A. Popp. 1983b. Mechanisms of formaldehyde toxicity. Pp. 132-147 in Formaldehyde Toxicity, J.E. Gibson, ed. New York: Hemisphere Publishing.

Thomson, E.J., S. Shackleton, and J.M. Harrington. 1984. Chromosome aberrations and sister-chromatid exchange frequencies in pathology staff occupationally exposed to formaldehyde. Mutat. Res. 141(2):89-93.

Titenko-Holland, N., A.J. Levine, M.T. Smith, P.J. Quintana, M. Boeniger, R. Hayes, A. Suruda, and P. Schulte. 1996. Quantification of epithelial cell micronuclei by fluorescence in situ hybridization (FISH) in mortuary science students exposed to formaldehyde. Mutat. Res. 371(3-4):237-248.

Tong, Z.M., J. Shi, J.S. Zhao, H. Yang, R.M. Jiang, Z. Kong, and Q. Sun. 2006. Analysis on genetic toxicity of formaldehyde on occupational exposure population [in Chinese]. Chin. J. Public Health 22(7):783-784.

Vargova, M., S. Janota, J. Karelova, M. Barancokova, and M. Sulcova. 1992. Analysis of the health risk of occupational exposure to formaldehyde using biological markers. Analusis 20(8):451-454.

Vasudeva, N., and C. Anand. 1996. Cytogenetic evaluation of medical students exposed to formaldehyde vapor in the gross anatomy dissection laboratory. J. Am. Coll. Health 44(4):177-179.

Viegas, S., C. Ladeira, C. Nunes, J. Malta-Vacas, M. Gomes, M. Brito, P. Mendonca, and J. Prista. 2010. Genotoxic effects in occupational exposure to formaldehyde: A study in anatomy and pathology laboratories and formaldehyde-resins production. J. Occup. Med. Toxicol. 5(1):25.

Viegas, S., C. Ladeira, M. Gomes, C. Nunes, M. Brito, and J. Prista. 2013. Exposure and genotoxicity assessment methodologies: The case of formaldehyde occupational exposure. Curr. Anal. Chem. 9(3):476-484.

Vock, E.H., W.K. Lutz, O. Ilinskaya, and S. Vamvakas. 1999. Discrimination between genotoxicity and cytotoxicity for the induction of DNA double-strand breaks in cells treated with aldehydes and diepoxides. Mutat. Res. 441(1):85-93.

Von Hippel, P.H., and K.Y. Wong. 1971. Dynamic aspects of native DNA structure: Kinetics of the formaldehyde reaction with calf thymus DNA. J. Mol. Biol. 61(3):587-613.

Wang, B., and D.D. Liu. 2006. Detection of formaldehyde induced developmental toxicity assessed with single cell gel electrophoresis [in Chinese]. Fen Zi Xi Bao Sheng Wu Xue Bao 39(5):462-466.

Wilkins, R.J., and H.D. Macleod. 1976. Formaldehyde induced DNA-protein crosslinks in *Escherichia coli*. Mutat. Res. 36(1):11-16.

Wong, V.C., H.L. Cash, J.L. Morse, S. Lu, and A. Zhitkovich. 2012. S-phase sensing of DNA-protein crosslinks triggers TopBP1-independent ATR activation and p53-mediated cell death by formaldehyde. Cell Cycle 11(13):2526-2537.

Yager, J.W., K.L. Cohn, R.C. Spear, J.M. Fisher, and L. Morse. 1986. Sister-chromatid exchanges in lymphocytes of anatomy students exposed to formaldehyde-embalming solution. Mutat. Res. 174(2):135-139.

Ye, X., W. Yan, H. Xie, M. Zhao, and C. Ying. 2005. Cytogenetic analysis of nasal mucosa cells and lymphocytes from high-level long-term formaldehyde exposed workers and low-level short-term exposed waiters. Mutat. Res. 588(1):22-27.

Ye, X., Z. Ji, C. Wei, C. McHale, S. Ding, R. Thomas, X. Yang, and L. Zhang. 2013. Inhaled formaldehyde induces DNA-protein crosslinks and oxidative stress in bone marrow and other distant organs of exposed mice. Environ. Mol. Mutagen. 54(9):705-718.

Ying, C.J., W.S. Yan, M.Y. Zhao, X.L. Ye, H. Xie, S.Y. Yin, and X.S. Zhu. 1997. Micronuclei in nasal mucosa, oral mucosa and lymphocytes in students exposed to formaldehyde vapor in anatomy class. Biomed. Environ. Sci. 10(4):451-455.

Ying, C.J., X.L. Ye, H. Xie, W.S. Yan, M.Y. Zhao, T. Xia, and S.Y. Yin. 1999. Lymphocyte subsets and sister-chromatid exchanges in the students exposed to formaldehyde vapor. Biomed. Environ. Sci. 12(2):88-94.

Yu, L.Q., S.F. Jiang, S.G. Leng, F.S. He, and Y.X. Zheng. 2005. Early genetic effects on workers occupationally exposed to formaldehyde [in Chinese]. Zhonghua Yu Fang Yi Xue Za Zhi 36(6):392-395.

Zeller, J., A. Ulrich, J.U. Mueller, C. Riegert, S. Neuss, T. Bruckner, G. Triebig, and G. Speit. 2011a. Is individual nasal sensitivity related to cellular metabolism of formaldehyde and susceptibility towards formaldehyde-induced genotoxicity? Mutat. Res. 723(1):11-17.

Zeller, J., S. Neuss, J.U. Mueller, S. Kühner, K. Holzmann, J. Högel, C. Klingmann, T. Bruckner, G. Triebig, and G. Speit. 2011b. Assessment of genotoxic effects and changes in gene expression in humans exposed to formaldehyde by inhalation under controlled conditions. Mutagenesis 26(4):555-561.

Zeller, J., J. Hogel, R. Linsenmeyer, C. Teller, and G. Speit. 2012. Investigations of potential susceptibility toward formaldehyde-induced genotoxicity. Arch. Toxicol. 86(9):1465-1473.

Zhang, L.P., X.J. Tang, N. Rothman, R. Vermeulen, Z. Ji, M. Shen, C. Qiu, W. Guo, S. Liu, B. Reiss, L.B. Freeman, Y. Ge, A.E. Hubbard, M. Hua, A. Blair, N. Galvan, X. Ruan, B.P. Alter, K.X. Xin, S. Li, L.E. Moore, S. Kim, Y. Xie, R.B. Hayes, M. Azuma, M. Hauptmann, J. Xiong, P. Stewart, L. Li, S.M. Rappaport, H. Huang, J.F. Fraumeni, Jr., M.T. Smith, and Q. Lan. 2010b. Occupational exposure to formaldehyde, hematotoxicity, and leukemia-specific chromosome changes in cultured myeloid progenitor cells. Cancer Epidemiol. Biomarkers Prev. 19(1):80-88.

Zhao, Q., L. Duan, Y. Liu, Z. Lu, O.Y. Qia, Y. Yan, and X. Yang. 2004. Effect of formaldehyde from artificial based boards on micronucleus of mice marrow polychromatic erythrocyte. J. Public Health Prev. Med. 15(6):18-20.

Zhitkovich, A., and M. Costa. 1992. A simple, sensitive assay to detect DNA-protein crosslinks in intact cells and in vivo. Carcinogenesis 13(8):1485-1489.

Zhong, W.G., and S. Que Hee. 2004a. Quantitation of normal and formaldehyde-modified deoxynucleosides by high-performance liquid chromatography/UV detection. Biomed. Chromatogr. 18(7):462-469.

Zhong, W.G., and S. Que Hee. 2004b. Formaldehyde-induced DNA adducts as biomarkers of in vitro human nasal epithelial cell exposure to formaldehyde. Mutat. Res. 563(1):13-24.

Zhong, W.G., and S. Que Hee. 2005. Comparison of UV, fluorescence, and electrochemical detectors for the analysis of formaldehyde-induced DNA adducts. J. Anal. Toxicol. 29(3):182-187.